Post-Truth

Post-Truth

Knowledge as a Power Game

Steve Fuller

ANTHEM PRESS

Anthem Press
An imprint of Wimbledon Publishing Company
www.anthempress.com

This edition first published in UK and USA 2018
by ANTHEM PRESS
75–76 Blackfriars Road, London SE1 8HA, UK
or PO Box 9779, London SW19 7ZG, UK
and
244 Madison Ave #116, New York, NY 10016, USA

British Library Cataloguing-in-Publication Data
A catalogue record for this book is available from the British Library.

ISBN-13: 978-1-78308-693-1 (Hbk)
ISBN-10: 1-78308-693-9 (Hbk)

ISBN-13: 978-1-78308-694-8 (Pbk)
ISBN-10: 1-78308-694-7 (Pbk)

This title is also available as an e-book.

This book is dedicated to the memory of the founder of 'scientific history', the ancient Greek historian Thucydides, who by today's standards would be regarded as a purveyor of 'fake news'.

CONTENTS

ACKNOWLEDGEMENTS

Among the people (aside from the publisher!) who had to suffer graciously while I stretched their patience as I wrote various texts that provide components of the argument in these pages, let me single out for special thanks: James Chase, Jim Collier, Alistair Duff, Joannah Duncan, Bob Frodeman, Inanna Hamati-Ataya, Jerry Hauser, Ilya Kasavin, Sharon Rider, Mikael Stenmark and Jack Stilgoe. I would also like to thank the British Sociological Association, the European Association for the Study of Science and Technology, the *Guardian* newspaper and London's Institute of Art and Ideas for hosting earlier versions of what I say here on their websites. Finally, I would like to acknowledge support of the Russian Science Foundation, project number 14-18-02227, 'Social Philosophy of Science', with which I am associated as a research fellow in the Russian Academy of Sciences, Institute of Philosophy.

Introduction

SCIENCE AND POLITICS IN A POST-TRUTH ERA: PARETO'S HIDDEN HAND

'Post-truth' may have been declared word of the year for 2016 by the *Oxford English Dictionary*, but the concept has been always with us in both politics and science – and in much deeper ways than those who decry its existence realize. A long memory is not required to see its roots in politics. Recall 2004's coinage of 'reality-based community' as an ironic counterpoint to George W. Bush's approach to foreign policy, especially after the start of the Iraq war. Nevertheless, it is interesting to see the exact dictionary definition of 'post-truth', including examples of usage:

> Relating to or denoting circumstances in which objective facts are less influential in shaping public opinion than appeals to emotion and personal belief:
>
> *'in this era of post-truth politics, it's easy to cherry-pick data and come to whatever conclusion you desire'*
>
> *'some commentators have observed that we are living in a post-truth age'*

This definition is clearly pejorative. Indeed, it is a post-truth definition of 'post-truth'. It is how those dominant in the relevant knowledge-and-power game want their opponents to be seen. In this context, the word 'emotion' is a bit of post-truth jargon that only serves to obscure the word's true function, which is to gain competitive advantage in some more or less well-defined field of play.

Those who are most resonant to our living in a 'post-truth' world believe that reality is fundamentally different from what most people think. This applies to both sides of today's 'post-truth' divide: the elite experts *and* the populist demagogues. Informing it all is Plato's view of the world, which Niccolò Machiavelli helpfully democratized in the Renaissance. It was then updated for the capitalist world by the political economist Vilfredo Pareto (1848–1923), one of sociology's forgotten founders, an inspiration to Benito Mussolini and the man who was still cast in my youth as the 'Marx of the Master Class', given his respectful treatment by such Cold War liberals as Talcott Parsons and Raymond Aron (Parsons 1937: chaps. 5–7; Aron 1967: chap. 2). If anyone deserves to be the patron saint of post-truth, it is Pareto.

For Pareto, what passes for social order is the result of the interplay of two sorts of elites, which he called, following Machiavelli, *lions* and *foxes*. Both species are post-truth merchants. The lions treat the status quo's understanding of the past as a reliable basis for moving into the future, whereas the foxes regard the status quo as possessing a corrupt understanding of the past that inhibits movement into a still better future. History consists in the endless circulation of these two temporal orientations: the 'inductive' and the 'counter-inductive', as epistemologists would say.

The *Oxford English Dictionary*'s definition of 'post-truth' speaks the lion's truth, which tries to create as much moral and epistemic distance as possible from whatever facsimile of the truth the fox might be peddling. Thus, the fox – but not the lion – is portrayed as distorting the facts and appealing to emotion. Yet, the lion's truth appears to the fox as simplistically straightforward and heavy-handed, little more than claims to entitlement often delivered in a fit of righteous indignation. Thus, the fox's strategy is to minimize the moral and epistemic distance between his own position and that of his leonine opponent, typically by revealing her unredeemed promises and rank hypocrisy.

Post-truth politics was laid bare in the 2016 US presidential campaign when the leonine Hillary Clinton, perhaps the most qualified person ever to run for the presidency, called half of Donald Trump's supporters 'a basket of deplorables' for trying to undermine the dominant 'progressive' agenda of the post-Cold War neo-liberal welfare state. In response, the foxy Trump, speaking on behalf of the Americans increasingly left behind by this same agenda, called the people fronting it 'corrupt' and 'crooked'.

But Trump meant something deeper, which goes to the heart of the post-truth condition. It came across in his campaign catchphrase: 'draining the swamp'. The entire Washington establishment – not only Clinton's Democrats but also the opposing Republican Party who nominated Trump as their candidate – was blamed for having staged a rigged game in which whoever was elected, the ensuing legislation would always benefit the political class, regardless of its consequences for the populace. In more leonine days, this was called 'bipartisanship' and it got the business of government done. Indeed, its sociological defenders had been trailing it as the 'end of ideology' for at least two generations (Bell 1960). It was supposed to be the game that beats all games. But Trump successfully showed that it was still just one more game. That's the post-truth condition in a nutshell.

In philosophical slang, the post-truth condition is all about *going meta*. You try to win not simply by playing by the rules but also by controlling what the rules are. The lion tries to win by keeping the rules as they are, and the fox tries by changing them. In a truth game, the lion's point of view is taken for granted without much thought: opponents contest each other according to agreed rules, and this initial agreement defines the nature of their opposition and the state of play at a given moment. Here the foxes are potentially disgruntled losers. In a post-truth game, the aim is to defeat your opponent in the full knowledge that the rules of the game might change. In that case the nature of your opposition could change in a way that might flip the advantage to your opponent. Here the foxes are always playing for the flip.

When Machiavelli said that effective rulers use always force sparingly, he was talking about trying to maintain the truth game. Lions should not have to roar because they could well lose if they actually need to back up the roar with action, something that history has repeatedly shown. The truth game works best if there is only the threat but not the actual show of force from the self-appointed upholders of truth. Thus, the lion's strategy is all about quashing the counterfactual imagination, the thought that things might turn out to be other than they are. This constitutes an exercise of what I call in these pages *modal power*. Plato tried something similar by proposing censorship, so that artists would know in advance that their productions would not be tolerated if they crossed a certain line of political correctness, as determined by the philosopher-king. (Plato would find much to admire in today's campus 'speech codes'.) That Machiavelli felt he had to be more direct about these matters is a moment in the history of democratization and, so I believe, human self-consciousness more generally.

Trump raises the stakes still higher when he invokes 'fake news', which questions the conventional liberal vehicles by which the truth/false distinction is reproduced in the American mass media, such as the *New York Times* or CNN. Here he has been aided tremendously by the advent of social media, in which newsfeeds such as the anti-establishment 'alt-right' Breitbart have managed to be streamed alongside and sometimes in place of 'mainstream' news media on Facebook pages. (Breitbart's former chief executive, Steve Bannon, was Trump's chief strategist during the 2016 campaign and his first year in office.) The result is that people are provided with either conflicting news accounts, which they are then forced to resolve for themselves, or simply the news account that corresponds to their revealed preferences as a social media user. In either case, they are rendered more confident to decide matters of truth for themselves.

A good sign that people have 'gone meta' in response to this pluralization of news outlets is the level of paranoia that erupts on social media upon announcing that a story has been sourced by, say, Breitbart or CNN, depending on one's default ideology. After all, as much as viewers of CNN hate their favourite channel being dubbed 'fake news', that is exactly what they think of Breitbart and its fellow traveller on television, Fox News. And so the playing field has been levelled. Everyone plays Trump's game. And the name of the game is tit for tat, which fills up an unprecedented part of the news today: for every fact-check that CNN does of Trump's tweets, which Fox News tends to promote and even embellish with obliging commentators, Fox News draws attention to all the good that Trump has been doing, which CNN ignores because it is preoccupied with trying to find grounds to remove Trump from office.

Although we live in a world of 'rolling 24/7' news, what I have just described does not turn out to be quite a fair fight. In an increasingly democratic society, attention spans remain just as limited as ever, but now people find being effectively treated like dupes to be not much better than being explicitly treated like idiots. All of this plays to Trump's side of the 'meta-argument', which is ultimately about sufficiently trusting people's capacity to judge for themselves matters of truth that you would allow them to live with the consequences in cases where their judgement turns out to have been in error, at least from the standpoint of their own welfare or advantage.

Now turn to science, and the situation is not so different. Pareto's lions acquire legitimacy from tradition, which in science is based on expertise rather than lineage or custom. Yet, like these earlier forms of legitimacy, expertise derives its authority from the cumulative weight of intergenerational experience. This is exactly what Thomas Kuhn (1970) meant by a 'paradigm' in science – a set of conventions by which knowledge builds in an orderly fashion to complete a certain world view established by a founding figure – say, Sir Isaac Newton or Charles Darwin. Each new piece of knowledge is anointed by a process of 'peer review'.

What makes Kuhn's account of science 'post-truth' is that 'truth' is no longer the arbiter of legitimate power but rather the mask of legitimacy that is worn by everyone in pursuit of power. It turns out that Kuhn spent his formative years at Harvard in the late 1930s when the local kingmaker, biochemist Lawrence Henderson, not only taught the first history of science courses but also convened an interdisciplinary 'Pareto Circle' to get the university's rising stars acquainted with the person he regarded as Karl Marx's only true rival (Barber 1970; Fuller 2000b: chap. 4). The fact that Henderson protégé and Kuhn mentor, Harvard president James Bryant Conant, was among the last to support war against the Nazis but among the first to propose use of the atomic bomb to end the war against Japan, reveals someone who had learned his Pareto well. Conant jumped ship only when absolutely necessary (the lion's move) and seized the opportunity when others remained in doubt (the fox's move) (Hershberg 1993: chap. 6).

The most interesting feature of Kuhn's narrative of how science progresses – and his has been the most influential of any such narrative for the past half-century – is his 'Orwellian' characterization of the understanding of the history of science that both professional scientists and the general public need to have (Kuhn 1970: 169). Here Kuhn is alluding to *1984*, in which the protagonist's job is to rewrite newspapers from the past to make it seem as if the government's current policy is where it had been heading all along. In this perpetually airbrushed version of history, the public never sees the U-turns, switches of allegiance and errors of judgement that might cause them to question the state's progressive narrative. Confidence in the status quo is maintained and new recruits are inspired to follow in its lead. Kuhn claimed that what applies to totalitarian *1984* also applies to science united under the spell of a paradigm.

But lying on the cutting room floor are the activities of the other set of elites, the foxes, to whom professional historians of science – who generally have no vested interest in science sticking to its official line – turn to, to find out find what really goes on behind the scenes of the drama that the lions are trying to stage. In today's politics of science, the foxes are known by a variety of names, ranging from 'mavericks' to 'social constructivists' to 'pseudoscientists'. They are characterized by dissent and unrest, thriving in a world of openness and opportunity.

The lions of the scientific establishment at first respond to the foxy dissenters in their midst by simply denying their epistemic standing – if not outright existence – rather than contesting them explicitly on common ground. Thus, no matter their sophistication, creationists, climate change sceptics and various New Agers are not merely wrong, but they are also 'bad' in a sense that allows epistemic failure to bleed into moral failure. Thus, whenever dissenters claim to be weighting the evidence differently, they are denounced as liars for not upholding the orthodoxy.

But that dismissive strategy is prone to hazards, especially when the orthodoxy itself cannot quite establish the truth it has been promising to secure for so long – and when this temporizing is then followed by requests for more public money for further research. Add to that the feats of fudging, complicating, backtracking and all round 'adhockery' that the orthodoxy must routinely perform to show that it is getting closer to the truth – as opposed to adopting the simpler strategy of changing course altogether. The dispassionate observer might well conclude that the lion's extremely loud roar belies its inability to defeat any challengers who might call its bluff.

For their part, foxes stress the present as an ecstatic moment in which there is everything to play for, what the ancient Greek sophists originally called *kairos*. This includes a decisive break with 'the past', which they know is fictionalized anyway, as in *1984*. Self-styled visionaries present themselves, like Galileo Galilei, as the first to see what is in plain sight. Expertise appears as a repository of corrupt judgement designed to suppress promising alternatives to already bankrupt positions. For Kuhn, the scientific foxes get the upper hand whenever cracks appear in the lions' smooth narrative, the persistent 'anomalies' that cannot be explained by the ruling paradigm.

However, the foxes have their own Achilles heel: they are strong in opposition but divisive in office. Kuhn's great opponent Karl Popper (1981) put a brave face on this feature, echoing Leon Trotsky in calling for a 'permanent revolution' in science. But if the field of play in science is opened to all comers, then the rules of the game itself might change to become unrecognizable. Few scientists nowadays deny the need to extend the public's sense of 'scientific citizenship', but equally few would have it morph into 'proletarian science', whereby the research agenda is dictated by popularly elected committees (Lecourt 1976). In this respect, scientists do not wish to cross a line comparable to the one politicians face between parliamentary and participatory democracy, in which they shift from serving the people's real interests to serving what the people think their interests are. As we shall see in Chapter 1, exactly that line was crossed in the case of Britain's 2016 referendum to leave the European Union (EU), or 'Brexit'.

As of this writing, more than twenty books have appeared with 'post-truth' in the title, all of which centre on the role that the *Oxford English Dictionary*'s pejorative sense of 'post-truth' played in the twin victories of Brexit and Trump. They invariably focus on 'fake news', usually in a way that is biased against the victors, as per the dictionary definition. This book does not aim to be a competitor to theirs. On the contrary, as this introduction has suggested, I take post-truth to be a deep feature of at least Western intellectual life, bringing together issues of politics, science and judgement in ways which established authorities have traditionally wished to be kept separate. Even if Trump is forced to resign or fails to achieve a second term in office, or Brexit is reversed in the eleventh hour (all of which are possible as of this writing), the post-truth condition will remain. With that in mind, let me briefly survey this book's contents.

Chapter 1 adopts a broad social epistemological approach to Brexit, whereby the United Kingdom's parliamentary foxes got more than what they bargained for by encouraging the public to think from first principles about the nation's continued membership in the EU. The British electorate's taste for direct democracy has not abated, notwithstanding the uncertain future into which Brexit plunges the nation, which both the foxes and the lions are feverishly trying to sort out.

Chapter 2 shows that philosophy, a discipline that likes to present itself as pre-eminently concerned with 'The Truth' has appreciated the post-truth perspective throughout its history, starting with Plato's Dialogues, in which the leonine Socrates always outsmarts his foxy Sophist opponents. Moreover, even within the modern 'analytic' school, long the establishment face of academic philosophy in the English-speaking world, there never has been any agreement on either the nature or the criteria of truth, though clearly some definitions of truth favour some modes of thought and forms of knowledge more than others. In that respect, philosophy remains post-truth all the way down.

Chapter 3 is concerned first with sociology, arguably the consummate post-truth science, given its preoccupation with how people manage to redefine themselves under changing conditions, most notably from the pre-modern to the modern – and arguably postmodern – world. Unfortunately, science and technology studies, the leading edge of the sociology of knowledge, has in recent years retreated from its original embrace of the post-truth condition, even though it remains in the best position to illuminate science's deeply game-like character.

Chapter 4 diagnoses academia's failure to fully exploit even its own knowledge base in terms of the grip that disciplines or 'paradigms' (in Kuhn's sense) have on enquirers. These social formations effectively channel academic effort along default lines of enquiry. Luckily, not all academically trained and interested parties reside in the academy. More specifically, there has been what I call a 'military-industrial will to knowledge' which targets what library science calls 'undiscovered public knowledge'. The chapter ends with a reflection on 'information overload' as a more general cultural context in which the problem of undiscovered public knowledge arises.

Chapter 5 deals with the post-truth phenomenon of 'customized science', which consists in idiosyncratic interpretations and appropriations of scientific knowledge that, to varying degrees, contradict the authority of expert scientists. It is a natural outgrowth of a world in which science is seen as increasingly relevant to people's lives, while the sources of information about science have extended well beyond the science lab and classroom. The result is what I have called 'Protscience', as people have come to personalize their understanding of science in the manner of the Bible during the Protestant Reformation.

Chapter 6 takes off from Max Weber's famed complementary speeches on politics and science as a 'vocation' to argue that they are really about the same topic, the former under the guise of the post-truth condition and the latter under the guise of the truth condition. Put bluntly, scientists aim to discern the rules of the game that politicians are keen to change to their advantage. This epitomizes the struggle for 'modal power', which is to say, control over what is possible. The writing and rewriting of history is perhaps the field where this struggle is most clearly played out.

Chapter 7 closes the book with a discussion of the epistemology of the future, forecasting, the ultimate playground for the post-truth imagination. This is less about how to predict the future correctly than how to make the most of whatever happens. On this basis, history has shown that players who begin as losers can end up winners, simply because they take better advantage of their current situation, even if it has resulted from a major setback. The chapter surveys several attitudes to the future, most notably 'adaptive preferences' and its policy extension, 'precipitatory governance', which advises that instead of trying to avoid catastrophe, one should plan as if it were going to happen, since the plans may succeed even if doomsday is avoided. The Internet was a Cold War innovation developed in just this spirit.

Chapter 1

BREXIT: POLITICAL EXPERTISE CONFRONTS THE WILL OF THE PEOPLE

Introduction

I should start by saying that I would be happy to reverse the course which the United Kingdom has taken since that fateful 52/48 decision on 23 June 2016 to exit the EU after more than 40 years of membership (hence 'Brexit'). Any path that would lead the United Kingdom back to the EU is fine with me: a parliamentary vote, another general election, a second referendum – you name it. But suppose Brexit is inevitable. My view then is that we should examine more closely – and even more charitably – what some of the more 'visionary' Brexiteers have been projecting. However, this is not as easy as it first sounds because their vision is a strange amalgam of populism and elitism, which when taken together threatens not only the sovereignty of Parliament, which has been much discussed in the media, but also the authority of expertise more generally. Such are the ways of the fox, in Pareto's terms. Here it is worth recalling that virtually all of UK academia, business leaders – including the Bank of England – and world politicians who expressed an opinion wanted to see the United Kingdom remain in the EU. (Russia was a notable exception.)

However, as we shall see, Brexit has turned out to be a poisoned chalice for the Brexiteers, who had not anticipated that the public would treat its new-found voice as though it were a sort of collectively manifested expertise of its own. I present the argument that follows in three parts. First, I consider Brexit in relation to my own long-standing anti-expertist approach to social episte-mology, which in many ways makes me a kindred spirit to the Brexiteers. Next, I turn to the struggle of parliamentary elites which eventuated in the win for Brexit, focusing on the Brexiteers' distinctive epistemic and ethical strategy with regard to public opinion. Finally, I consider the unforeseen emergence of a Rousseau-style 'general will' with regard to Brexit, which is where British democracy stands for the foreseeable future, ending on the role of academia – and specifically business schools – in the anti-expert revolution.

The Anti-expert Turn in Politics and Science

The topic of expertise is close to my heart because the version of 'social epistemology' that I have been developing over the past 30 years has stood out for its 'deconstructive' and 'demystifying' attitude towards expertise, which I originally dubbed 'cognitive authoritarianism' (Fuller 1988: chap. 12). As a philosopher of science who became a 'social constructivist' in the formative years of the field now known as 'science and technology studies' (STS), I differed from my philosophical colleagues in seeing the disciplinary boundaries by which expertise is institutionalized as mere necessary evils vis-à-vis free enquiry: the more necessary, the more evil (Fuller and Collier 2004: chap. 2). In this context, I stood with Karl Popper as against Thomas Kuhn: the former said that no scientific knowledge claim is irreversible, the latter that science depends on its knowledge claims being rarely reversed (Fuller 2003a).

When I turned to 'knowledge management' about 20 years ago, I was struck by the Janus-faced way in which economics portrayed knowledge in wealth creation. On the one hand, it appeared as a magic 'X factor' in the production function, usually called 'innovation', which is irreducible to the available epistemic and material resources. On the other hand, there is knowledge as 'expertise', a form of *rent-seeking* that is structured around having to acquire credentials before accessing what is already known (McKenzie and Tullock 2012: part 5). It was Popper and Kuhn all over again. From the standpoint of a dynamic capitalist economy, innovation is clearly positive, not least because it 'creatively destroyed' markets, the functional equivalent of a paradigm shift in science. In contrast, expertise is seen negatively as a major source of information bottlenecks. At the time, I believed that the emergence of 'expert systems', whereby computers are programmed to reproduce the reasoning of experts under normal conditions, might ultimately remove such bottlenecks by rendering human experts redundant, not least in relatively high-paying but routinized fields of law and medicine. That future is still very much on the agenda (Fuller 2002: chap. 3).

Still more recently, I have become concerned about the future of the increasingly 'research-led' university, which is arguably a euphemism for the institution's role in the manufacture and certification of expertise. In this spirit, I have called for a shift in the university's mission from research back to teaching, which has historically done the most to break down the hierarchies, or 'bottlenecks', that expertise breeds (Fuller 2016a). In this context, teaching should be seen as the regular delivery of knowledge to those who would otherwise remain ignorant by virtue of being removed from the channels in which such knowledge normally travels. To be sure, this levelling of epistemic authority enables more people to 'own' formerly expert knowledge, in the resonant sense that 'own' enjoys today. But at the same time, it removes the stabilizing effect that expert knowledge has had on the social order in the past, given that a wider range of people can take the same knowledge in a wider range of directions.

Arguably, this collective epistemic volatility has been intensified in our own day with the rise of the Internet as society's principal means of knowledge acquisition. And just as the Protestant Reformers 500 years ago capitalized on the advent of the printing press to delegitimize the authority of the Roman Catholic Church by urging the faithful to read the Bible for themselves, various anti-establishment campaigners in both politics and science have urged their followers to override the experts and judge the evidence for themselves.

I have never seen much of a difference between the epistemologies of politics and science. Here I stand closer to Karl Popper than to Max Weber, two thinkers who otherwise share many similar sensibilities. As someone who has been intimately involved with one of the major anti-expert science movements in our time, *intelligent design theory*, I see some striking similarities with Brexit. Intelligent design theory is a form of scientific creationism that is premised on the idea that life is too complexly specified to have been the product of unintelligent variation and selection processes, à la Charles Darwin's theory of evolution (Fuller 2007a; Fuller 2008).

The first and perhaps most important similarity is that an institutional opening already existed for the experts to be challenged. In the case of intelligent design, it was built into the US Constitution, namely, the devolution of education policy to the local tax base, which funds the school system. The original idea was to prevent education from being dominated by the secular equivalent of an established church, or a 'national religion'. In that context, academic authorities function no more than as consultants and lobbyists in terms of curriculum construction and textbook purchases, which are ultimately in the hands of local school districts. In the case of Brexit, the opening was provided by Parliament's right to call a referendum, thereby throwing open to a direct public vote what would otherwise be a statutory issue. This right has been rarely exercised in Parliament's long history. Moreover, unlike the United States, where the referendum is commonly used by several states to determine matters such as setting tax rates, on which voters might be expected to have relatively well-formed views, the United Kingdom has called a referendum only on relatively esoteric high-level matters of governance, such as proportional representation and, of course, membership in the EU.

To be sure, intelligent design theory has been hoist by its own petard in US courtrooms, as it is regularly ruled to be a crypto-Christian plot to overturn secular democracy. Yet, there is little evidence that the well-publicized legal defeats suffered by the theory have diminished public support for it. Perhaps more to the point, there is equally little evidence that these defeats have served to increase the public's belief in evolution, let alone public trust in the scientific establishment that backs evolution. Instead there is a climate of suspicion and even paranoia that agencies of the state are on a mission to subvert dissenting voices that uphold Christian values.

Indeed, if evolution were subject to a national referendum in the United States, it might well lose by something like a Brexit-style 52/48 margin. Trump, playing somewhat against type, managed to capitalize on that sentiment in his path to the White House. Similarly, even after the triumph of Brexit at the ballot box, there is widespread scepticism that it will be implemented in the spirit of the referendum campaign, given that the House of Commons was 4 to 1 – and the Lords 6 to 1 – in favour of remaining in the EU. And while the numbers in the Commons have shifted towards Brexit as a result of the 2017 general election, parliamentarians generally want to remain as close as possible to the current UK-EU arrangement rather than what the public seemed to have wanted, namely, to reboot Britain's place in the world. A reasonable inference is that, for better or worse, the public is much less risk averse than its elected representatives.

The other important factor in the anti-expert revolt common to intelligent design and Brexit is the establishment's own admission that there are problems, but that these can be solved by staying within status quo. Where intelligent design theory goes beyond earlier forms of Creationism is that it not only argues for an alternative basis for explaining the nature of life (i.e. an 'intelligent designer', also known as the Abrahamic deity) but also addresses issues that evolutionists have already identified as problematic for their own account. Similarly, and perhaps fatally, Prime Minister David Cameron started the campaign to remain in the EU by conceding the EU's shortcomings, a panto version of which had been enacted in an ineffectual February 2016 Brussels summit, yet he also argued that these will not be remedied unless the United Kingdom stays to reform the EU from within.

Over time this message morphed into what Brexit campaigners dubbed 'Project Fear', namely, a generalized foreboding about the calamities that would follow from United Kingdom leaving an 'always already' flawed EU. Likewise, as support for intelligent design theory increased, the scientific establishment amplified the theory's threat to encompass all of science, if not civilization as such, were it to be taught in schools. Once again, on both matters, the public appears to be much less risk averse than the experts. But equally, by conceding fallibility at the outset the experts unwittingly opened the door to the public taking matters into its own hands.

At this point, we confront one of the big canards perpetrated by defenders of expertise: namely, that anti-experts are anti-intellectuals who privilege ignorance over knowledge and would treat all opinions as equally valid. All that this exercise in misdirection does is to cover up the reverse tendency, namely, that our trust in experts in modern democracies has led to a moral dumbing down of the population, as people are encouraged to let authorized others – starting perhaps with the general medical practitioner – decide for them what to believe, even when the consequences of those decisions directly affect people's lives and sense of self. In effect, modern democracy presents a paradox. At the same time as we enfranchise more of humanity into the political system, and indeed provide people with the education needed to function in that system, we are also discouraging them from exercising their judgement, given the increasing normative weight invested in expertise. The result is that we are breeding a culture of intellectual deference, a 'soft authoritarianism', if you will, whereby education ends up functioning in a counter-Enlightenment manner. Instead of individuals learning how to expand their powers over themselves and the wider world, they are being taught simply to discover and respect the limits of those powers.

What is missing here – and which Brexit's anti-expertism aims to inspire – is an ethic of intelligent risk-taking, one that acknowledges the full complexity of the world, which in turn requires diverse forms of knowledge, each of which is inherently partial and fallible. Moreover, given the pervasive unlikelihood that a perfect outcome can ever be reached, a democracy should aim to take decisions for which the people to whom they apply would be willing and able to take personal responsibility, whatever the consequences. In Kantian terms, all legislation should aspire to be self-legislation. Put more practically, reducing the difference between what the Parliamentarian and what the public knows about matters relating to the public good should be an aim of electoral politics.

I have associated this mindset with a *proactionary* – as opposed to a *precautionary* – approach to decision-making with regard to the future of the human condition (Fuller and Lipinska 2014). Clearly, putting the electorate in this mindset takes some persuasion in contemporary democracies, where voters often quickly punish politicians who fail to deliver on promises for which the voters have only themselves to blame. Interestingly, the Brexit-voting public so far has not engaged in this strategic distancing from its own decisions. If anything, as we shall see, it suffers from the reverse problem: The public has insisted that politicians get on with implementing the 'will of the people' in the case of Brexit, however ill-prepared, incoherent or potentially disastrous such a policy may be.

How the Anti-experts Beat the Experts at Their Own Game in Brexit

It is somewhat ironic that Brexit has propelled the current anti-expert revolution, given that it resulted from a referendum that was itself the product of infighting among parliamentary elites – indeed, within the same ruling party. However, the situation would have been all too familiar to Pareto, who in the introduction we highlighted as having seen the lifeblood of society as fuelled by the circulation of elites, whom he divided à la Machiavelli into 'lions' and 'foxes'.

In the case of the Brexit referendum, the lions were represented by those wanting to remain in the EU, including the Conservative prime minister, Cameron, who ill-fatedly called the referendum to stave off the foxes, as represented by two of his cabinet ministers and potential leadership rivals, Justice Minister Michael Gove and Minister without Portfolio Boris Johnson. A more specifically British analysis of this divide would start with the famous 'Two Cultures' problem that C. P. Snow first identified in 1956, as the reins of government were passing from the 'humanists' to the 'scientists' (including social scientists, especially economists and policy-oriented sociologists) with the advent of the post-war welfare state. In this respect, Brexit marked the ironic revenge of the humanists, given Gove's and Johnson's literary education, which neither has ever attempted to mask.

In terms of Snow's two cultures, the relevant division in the Brexit referendum was between those who tried to gain power by conjuring up verbal images of a much better world within reach ('humanists') by breaking with the EU and those who tried by brandishing statistics about how remaining within the EU was responsible for the rather good world in which people already live ('scientists'). Thus, whereas the 'foxy' Gove and Johnson stoked up passions in their respective columns for the *Times* and the *Telegraph*, the 'leonine' Cameron relied on sober economic arguments and forecasts with the backing of the Bank of England. And in that round of Snow's culture war, the rhetoricians ended up vanquishing the technocrats.

To be sure, what followed has not exactly gone to plan for the anti-expertists: Neither Gove nor Johnson is prime minister, and their Conservative Party is severely weakened in Parliament, following the 2017 general election. However, it is worth dwelling on how the Brexiteers outsmarted their opponents by effectively approaching the business of democracy from a rather different epistemological and perhaps even ethical standpoint.

The default position of public opinion research, as captured by most polling and survey work, targets a representative sample of an entire population according to such presumptively relevant variables as race, class, gender, age and so forth. The subjects are asked a series of questions, including their likely vote in the relevant contest. While this approach seems scrupulously fair and democratic, it nevertheless falls hostage to many possible interaction effects between researchers and subjects, not least ones involving their perceived social distance from one another, which can influence the subjects' candour. Thus, the recent spate of psephological failures (i.e. failures to predict voter behaviour), from Brexit to Trump, typically underestimated the 'populist' side of the argument. That was at least partly due to subjects not being entirely comfortable exposing their own train of thought, given what they imagined the researcher's to be. In other words, notwithstanding the methodological care that goes into preparing surveys and targeting populations, in the end subjects frame the situation as a task to appear respectable to a potentially judgemental stranger. The accuracy of what they say is probably not in the forefront of their minds, which in turn renders the researcher susceptible to misdirection when interpreting subjects' responses.

The anti-expert strain in Brexit and related campaigns had no trouble seeing the problems here and proceeded in a completely different way, which was not only more effective but perhaps also more democratic and fairer to the people concerned. In the first place, the anti-expertists do not treat the public as some radical 'other' from the pollster, whereby the former is full of 'attitudes' for which the latter then is a neutral sounding board, perhaps in the manner of a psychoanalyst who enables a client to unleash her unconscious. On the contrary, the anti-expertists are, if nothing else, social constructivists who appreciate the manufactured character of psephological knowledge.

Moreover, the classic self-understanding of public opinion researchers as neutral sounding boards no longer makes sense, now that the public has lived with polling and surveys for nearly a century. On the one hand, the work done with such instruments has become so routinized that its potential social distancing effects are easy to spot even by the 'lay' public; on the other, the public has got used to both the sort of questions that they are likely to be asked and the various effects that polls and surveys tend to have on the outcome that is of interest to the researcher. In short, through repeated exposure, the public has effectively acquired an understanding of the shortcomings which marketers have long associated with relying on 'stated' rather than 'revealed' preferences when trying to anticipate consumer behaviour.

In addition, to the anti-expertist, psephological knowledge is not only manufactured but also softly coercive, insofar as polls and surveys provide subjects with a wider range of options than perhaps they have ever explicitly considered. It rarely occurs to the public opinion researcher that subjects' responses may have less to do with bringing to the surface latent psychic tendencies than with simply generating behaviour that appears appropriate to the question being asked. To avoid such pitfalls, anti-experts favour techniques that owe more to psychological warfare than conventional public opinion research. In particular, Cambridge Analytica, a US/UK data analytics firm founded by Silicon Valley mogul and early Trump backer, Robert Mercer, has garnered attention for allegedly having done more to turn the United Kingdom to Brexit and the United States to Trump than anything that might have been done by any Russian hackers (Cadwalladr 2017).

A data analytics firm classifies and correlates a target population across many dimensions based on revealed preferences from a variety of sources, including census data, consumer purchases and prior electoral behaviour – all compiled with the latest computer technology alongside the user data routinely gathered by such social media platforms as Facebook. On this basis, it is possible, even without formally interacting with any subjects, to infer how many, if not most, of the population is likely to vote on a given issue, as long as opinion on the issue divides along these pre-existent preferences. To be sure, on some issues, segments of the population may display preferences that could appeal to either side of an issue. Public opinion researchers call these people 'swing voters', and on a very broad issue like Brexit the nature of the 'swing' is itself 'swingable', depending on how the issue is actually framed. Brexit is ripe for such treatment because the United Kingdom's role in the EU is an issue of quite broad policy scope, and voters generally lack a well-developed understanding of how it all matters to their lives.

The task for the campaign strategist then is to frame the issue – in this case Brexit – so that a relatively limited number of voters are up for grabs in the relevant swing and the campaign can stay focussed. Put another way, the worst-case scenario is one in which the issue is framed in a way that forces too many voters to rethink their preferences from scratch, as might have happened, had the Brexiteers focussed primarily on political corruption and wasteful expenditure at the EU. While that might have given the entire campaign a more 'Protestant Reformation' feel, it would also have left the outcome too much to chance, given that the charge could be thrown back at UK Parliament, in light of its own recent expenses scandal. Thus, specific talking points, such as '£350 million per week for the NHS' (National Health Service), came to epitomize the campaign, since – sticking to that example – it created the impression of a trade-off between funding the EU and national healthcare provision, especially in a time of fiscal austerity. While the figure and the reasoning behind it were called out during the campaign, it nevertheless succeeded in getting just the right people – the targeted swing voters – to personalize the trade-off between the interests of the EU and the United Kingdom.

Interestingly, Carole Cadwalladr (2017) and others present this anti-expertist use of data analytics in the spirit of an exposé, yet the supposedly exposed principals continue to act as if they have nothing to hide. Indeed, what their critics would identify as 'shamelessness' appears to be their rhetorical calling card. But do the anti-experts really have anything to be ashamed of? In our post-truth condition, the answer is no – on two grounds, one relatively technical and the other more broadly philosophical.

First, a distinctive feature of this advanced data analytics approach is that it does not coerce people to state any preferences. On the contrary, its 'unobtrusive methods' respect whatever preferences the voters have revealed in their behaviour. At most, the approach can be accused of 'nudging' voters in directions to which they were already inclined. In any case, it is up to the voters themselves to decide what to make of whatever campaign pitches appear on the newsfeeds they read. After all, as was noted above, in the near century since the advent of marketing and public opinion research, people have become accustomed to its character across a broad range of media. It would be patronizing to suggest that people are unaware of the presence of such efforts, even as they behave in accordance with them. This is not to deny that there is room for the public to improve its 'media literacy', but that requires personal trial and error just as much as any formal training, especially given a rapidly changing media landscape.

Moreover, this experimental approach to public opinion is prima facie in accord with Popper's vision of democracy as the 'open society', which amounts to turning society into a living laboratory. The anti-experts want to expedite this process, given the increased powers in data generation, surveillance and feedback afforded by advanced computer technology. The resulting social order is what one of today's leading futurists, Parag Khanna (2017), calls 'direct technocracy'. To its champions in the 'Hard' or 'Clean' Brexit camp, it casts the future United Kingdom as 'Switzerpore', combining the best of Switzerland and Singapore on the global stage. To its more dystopically inclined detractors, direct technocracy amounts to a reality television approach to policymaking that resembles a cross between *A Clockwork Orange* and *The Truman Show* (Kane 2016).

But at a deeper epistemological level, the horror displayed in Cadwalladr (2017) is symptomatic of a failure to recognize that even before Brexit's high-tech machinations, the 'facts' have always existed in the state of scare quotes, not only in politics but also in science. We see this virtually every day in the wrangling over the meaning of supposedly 'hard data' relating to inflation, unemployment, public expenditure, tax revenue, income levels, stock market activity and trade receipts. In these matters, it is not at all clear that a Nobel Prize winner and former chief economist at the World Bank such as Joseph Stiglitz has any special rhetorical advantage over a policy analyst placing a clear partisan spin on the figures.

This relatively levelled playing field does not reflect a lack of respect for the authority of science, but a recognition that scientific facts are 'hard' only in the context of academically defined games of 'hypothesis testing', in terms of which the contestants may gain or lose plausibility. Outside of that context, such 'facts' function more as placeholders, perhaps even metaphors, for a desired direction of policy travel. It matters less what the exact numbers are than that they are going in the right direction. Thus, Brexit campaigners sleep soundly even if '£350 million per week' is far from the likely sum that the UK government will have to spend on the NHS, since (so the campaigners argue) the voters ultimately just want to see a more substantial sum regularly spent. Similarly, the Bank of England economists who predicted incorrectly that the UK economy would collapse soon after the Brexit vote still feel vindicated by the pronounced slowing of economic growth in the ensuring period. Thus, each side continues to think that it has got hold of the truth, while the opponent is criticized for purveying 'fake facts'.

How the Anti-experts Ended Up Scoring an Own Goal in the Brexit Game

The fly in the ointment is that the people who are subject to the data analytic treatment just described do not seem to share its Popperian experimental attitude. It is this prospect that now worries the anti-expertists – and should worry us all. The data analytic strategists supporting Brexit thought that they had got round this problem by primarily working with revealed rather than stated preferences, thereby circumventing any self-masking by subjects that might make them appear more pro-EU than they really were. However, Brexiteers had not anticipated that people would so strongly identify with the referendum outcome as to make it difficult to reverse. But as it happens, the argument that was probably most persuasive with those who voted for Brexit had nothing to do with specific policy promises. Were that the case, one might see support for Brexit ebb away as the promises were shown to be undeliverable – or at least not deliverable in the relatively cost-free way presented during the campaign. There might then be an appetite for a parliamentary reversal of its commitment to Brexit, if not a second referendum. However, as we enter the second year of Brexit negotiations, there are few signs of this happening, which suggests that what really sold Brexit was the perceived return of *popular sovereignty*, something that is understood to be a good in itself, regardless of its actual exercise and consequences for the people to whom it was sold. In effect, the line between thinking and being in politics was erased.

Here one comes to appreciate a great virtue of parliamentary democracy: because Members of Parliament are officially 'representatives' of the people, they normally have the rhetorical space to reinterpret the 'public good' in light of changing circumstances without ever having to suggest that the people's own conception of the public good might be in error. Indeed, Members accept that they are the potential scapegoats for any perceived policy failures, as regular elections provide the populace with the opportunity to blame and replace them, whilst the populace itself remains shielded from having to take personal responsibility for the state of the nation. A binding referendum removes that convenient fiction, as the people apply their collective fingerprint to their fate.

Ironically, the anti-EU elites who favoured holding the referendum, largely to turn Britain into a proving ground for novel trade arrangements and other economic schemes (so-called 'Hard' or 'Clean' Brexit), managed instead to unleash the public's taste for Jean-Jacques Rousseau's 'general will'. This phrase, which became a slogan in the French Revolution, postulates the infallibility of a collective that is bound together by common values and experience. The contrasting notion, the 'aggregate will', is the sort of false ideal that Jean-Jacques Rousseau might have ascribed to the EU, at least as seen by its critics, in which the preferences of disparately interested parties are simply added together to reach policy decisions. What linked the collective will with infallibility in Rousseau's mind was the shared sense of identity that is performed by the many acting as one: to disagree with me is no longer simply to challenge my disposable opinion; rather, it is to challenge my very sense of who I am – in this case, a 'Briton'. It is worth recalling that a feature common to Conservative and Labour Party supporters who backed Brexit was a strong sense of the nation's identity being under threat, be it from foreign migrants or liberal cosmopolitans (Johnson 2017).

That Brexiteers realize that their work is cut out was revealed in a recent Twitter exchange reported by David Allen Green, legal commentator for the *Financial Times*, who blogs under the moniker 'Jack of Kent' (Green 2017). Green's interlocutor was one of the most 'shameless' of anti-experts, Dominic Cummings, who first rose to prominence as special adviser to Michael Gove during his stint as UK Secretary of State for Education and then went on to become the chief strategist for the successful Brexit campaign, which Gove co-chaired. When Green and others asked Cummings whether he thought – more than a year after the vote to leave the EU – that the British people would be happier under Brexit, he responded that government strategy now should be to maximize Britain's 'adaptiveness' to a broad range of possible futures, each of which involves its own form of happiness. What this probably means in practice is that the potential gains of leaving the EU should be promoted whilst downplaying the costs, even if the gains cannot quite make up for the costs in the way 'gains' and 'costs' were understood on the referendum day.

A test case for this strategy is the proposal that the United Kingdom post-Brexit will be open to all sorts of free trade deals around the world that it cannot make while remaining in the EU. That this may entail an immediate loss of considerable trade from Europe without equally immediate compensation from trade with other parts of the world could be portrayed as an 'investment' in a long-term vision of Britain as the world's major power that is most open to trade from anywhere in the world. But, of course, this 'investment' will affect various sectors of the population differently, and so one needs not only to distribute the costs equitably but also to adapt the presentations of those costs accordingly, so that people do not think that their lives are being made too miserable for too long. In any case, the goal remains one of protecting people from ever having to admit they made the wrong decision in the first place. But is this rather contorted and arguably quite manipulative course of action – simply to maintain the illusion of the general will's infallibility – really necessary?

This would involve demonstrating that voting for Brexit had been a mistake. The demonstration would take two steps. First, former prime minister Cameron, who resigned immediately after the vote, would admit that he called the referendum mainly to resolve a long-standing internal dispute within his own ruling Conservative Party, the festering of which was fuelling the electoral fortunes of the UK Independence Party. This strategy made sense to Cameron at the time because he already had form in calling two referenda – one on Scottish independence and another on an alternative voting scheme – and winning them in ways that sidelined his opponents. However, Cameron underestimated the ease with which the EU could be scapegoated for any number of problems that weighed heavily on the public's mind, which made Brexit look like a neat one-stop solution. This brings us to the second and decidedly more difficult step, which would involve telling the electorate that after being set up to vote for politically self-serving reasons, they managed to make the wrong choice. Whether or not the public was deceived is moot at this point. The fact remains that nine out of the ten UK regions that benefit the most from EU funding voted for Brexit. The harsh verdict of 'Turkeys voting for Christmas' is not unreasonable under the circumstances.

The question, then, is whether UK parliamentary democracy would be able to survive this double admission of error. Apparently, no one – other than former Tory Chancellor Ken Clarke, now the longest-serving parliamentarian, and Liberal Democrat leader Vince Cable – seems to want to chance it. Perhaps this reluctance for politicians of all parties to admit error is related to the United Kingdom's lack of a written constitution. After all, the casualness with which a binding referendum of such enormous national import as Brexit was enacted could be matched by another binding referendum that serves to fatally restrict the powers of Parliament itself, were it to concede that the Brexit vote was a gross misjudgement. In the end, the anti-expertist elites like Gove and Johnson, who managed to beat Cameron at his own game, failed to realize that members of the public might take the calling of the referendum to imply that they themselves possess their own kind of expertise.

It is worth recalling that in the middle third of the twentieth century, such promoters of an expert-steered mass society as Walter Lippmann (1922) and Alfred Schutz (1946) were concerned that the increasingly sensory character of the news media – whereby print is supplemented if not replaced by sound and vision – would result in people acquiring a kind of 'pseudo-experience' that would lead them to think that they know more than they really do, simply because they can claim to have 'seen' or 'heard' certain things broadcast in the media, which is then mixed with genuine personal experience to generate what they regard as politically valid judgements. And indeed, 'expert' is ety-mologically a contraction of 'experienced'. The original 'experts' were people who could demonstrate in a court of law that they had previously witnessed a pattern of behaviour that is relevant for deciding a case (Fuller 2002: chap. 3). It was also assumed that the significance attached to that experience was 'reli-able', which carried a double meaning of 'repeatable' and 'trustworthy'. This implied issues of training and accreditation, which turned 'expertise' into a high-rent epistemic district, whereby the significance of what one had 'experi-enced' in this sense was *both* inviolate and circumscribed.

The resulting image – promoted by Lippmann, Schutz and others – was that social order in complex democracies requires a 'distribution of knowledge' or 'division of cognitive labour'. The proposal basically generalizes the feudal model behind the cartographic imaginary that continues to lead academics to characterize their expertise as 'fields' and 'domains' of knowledge, bounded by rituals of mutual respect and deference. UK parliamentary politics also largely presupposes this model, which has created a class of 'professional poli-ticians', elected representatives – not delegates – of the people, whose public service ethos is bred from an early age in the independent schools that the British have traditionally called 'public'. To be sure, the anti-expertists are officially opposed to this model, notwithstanding their own roots in its elitist sensibilities. But they are one with the experts in sharing the same ultimate foe – the Rousseauian idea that a special wisdom is to be found in crowds, especially when they are agreed on a course of action, even if it remains radi-cally unclear how it might be brought about and what its outcomes might look like. Historically, such a gap between will and knowledge has been often made up by demagoguery, whereby one dictatorial figure comes to personify the publicly voiced aspiration. While it is highly unlikely that the United Kingdom will go down the demagogic route, the nation's next few months and years will constitute a very interesting high-risk experiment in democracy with much to teach the world, whatever the outcome.

Even though the anti-expert revolution has not quite gone to plan, it remains the direction of travel in democracy's fitful progress. Moreover, academia is clearly in its sights. My own antipathy to expertise has been inspired by Joseph Schumpeter's account of entrepreneurial innovation as 'creative destruction'. Ever since at least Fuller (2003a), I have been arguing that the university needs to apply Schumpeter's lesson to itself to avoid being seen as mainly in the business of creating bottlenecks to the flow of knowledge by dispensing credentials and engaging in other 'gatekeeping' activities that to the jaundiced eye look like rent-seeking. (Note the roots of the gatekeeping metaphor in medieval toll charges.) In this spirit, I have called for the university to relaunch its Enlightenment mission by stop privileging research activity, which arguably lays the groundwork for rent-seeking through priority claims made in the scholarly literature and the patent office. The academic capacity for countering expertization is expressed in the classroom, as teaching provides access to knowledge to those who would otherwise not acquire it because they have not been part of the contexts in which such knowledge is produced and distributed. Thus, universities manufacture knowledge as a public good through the creative destruction of the social capital formed by research networks. That is their unique selling point (Fuller 2009; 2016a).

Business schools may be uniquely positioned to perform this function. Their faculty are themselves often trained outside the academic field of business and, if the business schools are any good, most of the people they train will not remain in the academic field of business. If any part of the university deserves to carry the torch for anti-expertism, it is business schools. Indeed, if I had to organize a philosophy department, I would populate it with those who for the most part were not originally trained in the field, who are then entrusted with instructing a student body who for the most part intend to work outside of the field. In other words, it would look like a contemporary business school. It might even flourish as well as business schools have, if it adopted that organizational model. However, my point is about the aptness of the model to the discipline of philosophy, rather than its exact financial consequences. Rather like the word 'business', 'philosophy' would then primarily refer to the forum – or market (the same word in Greek, *agora*) – where people who would otherwise be strangers to each other are given a free space to transact knowledge claims, presumably to the mutual benefit of both, perhaps even to achieve a whole greater than the sum of its parts. In principle, this space could be anywhere, but the classroom would be an exemplary space for such transactions. The overall aim of the exercise would be to generate a common sense of humanity.

To be sure, the transaction is not exactly symmetrical. No goods are sold unless there is something to sell, and so the academics must be first to step up to the plate, trying to persuade prospective students that there is something they need to know. At a more abstract level, this opening gambit underwrites the idea of burden of proof, the normative foundation on which any real world argument rests: if I claim that you are lacking, then the burden is on me to prove it. It follows that I must possess some goods worthy of delivery. This is shown by what the students take away from what the academics have offered. Inspired by Adam Smith, the economist Deirdre McCloskey (1982) has long argued for a similarity – if not identity – between negotiating ideas and goods, which she associates (rightly, I believe) with the art of rhetoric. In that case, what passes for the 'canon' of Western philosophy may be understood as the sort of knowledge that is forged in such a spirit of supply and demand. On the pages of canonical works, the reader will find many real and imagined transactions being conducted and sometimes even resolved. Plato's dialogues are only the most obvious case in point, which perhaps explains why he always opens the canons, to which the rest of philosophy is 'footnotes', at least according to Alfred North Whitehead, who was not exactly Plato's biggest fan. In any case, works other than the ones currently venerated could have been selected to elicit a comparably 'canonical' effect – and probably will be selected in the future.

Chapter 2

WHAT PHILOSOPHY DOES AND DOES NOT TEACH US ABOUT THE POST-TRUTH CONDITION

A Post-Truth History of Truth

Philosophers claim to be seekers of the truth, but the matter is not quite so straightforward. Another way to see philosophers is as the ultimate experts in a post-truth world. They see 'truth' for what it is: the name of a brand ever in need of a product which everyone is compelled to buy. This helps explain why philosophers are most confident appealing to 'The Truth' when they are trying to persuade non-philosophers, be they in courtrooms or classrooms. In more technical terms, 'truth' – and the concepts surrounding it – is 'essentially contested' (Gallie 1956). In other words, it is not simply that philosophers disagree on which propositions are 'true' or 'false' but more importantly they disagree on what it means to say that something is 'true' or 'false'.

If you find my judgement too harsh or cynical, consider the careers of the key philosophical terms in which knowledge claims are transacted, not least 'evidence' and 'truth' itself. 'Evidence' is a good place to start because it feeds directly into the popular image of our post-truth world as 'post-fact', understood as a wilful denial of solid, if not incontrovertible, pieces of evidence, whose independent standing sets limits on what can be justifiably asserted about the world.

Yet it was only in the early modern period that philosophers even began to distinguish a purely fact-based conception of evidence from personal revelation and authoritative testimony. The break only became clean in the mid-nineteenth century when logic books regularly started to classify people-based claims to evidence among the 'informal fallacies', unless the people had direct acquaintance with the specific matter under dispute (Hamblin 1970). The concept of 'expert', a late nineteenth-century juridical innovation based on a contraction of the participle 'experienced', extended the idea of 'direct acquaintance' to include people with a specific training by virtue of which they are licensed to inductively generalize from their past experience to the matter under dispute. In this way, the recently proscribed 'argument from authority' made its return through the back door (Turner 2003).

This slow crafting of the concept of evidence was part of the general secularization of knowledge. At the same time, it would be a mistake to think that today's concept was purpose-made for scientific enquiry. Rather, it was an adaptation of the inquisition, the procedure used on the European continent to identify heretics and witches. Its English importer was Francis Bacon, King James I's lawyer, who believed that nature itself was a fugitive from the law, hiding its secrets from humanity for much too long. Special trials were thus required to force nature from its normally equivocal stance to decide between two mutually exclusive options (Fuller 2017).

Bacon called such trials 'crucial experiments', which Karl Popper turned into the gold standard of the scientific method three centuries later. To be sure, Bacon and Popper were under no illusions that the facts produced under such 'extraordinary rendition', as we would now say, were nature's deliverances in more relaxed settings. On the contrary, Popper went so far as to call facts 'conventions', by which he meant convenient way stations in a never-ending inquisition of nature. After all, what made experiments 'crucial' was that their outcomes hastened knowledge of a future that otherwise would only unfold – for good or ill – on nature's timetable, which would provide humanity little opportunity to plan a response, let alone steer nature's course to human advantage.

As for 'truth', it harks back to an older English word, 'troth', which harbours all of the concept's philosophical difficulties. 'Troth' means faithfulness – but to what exactly: the *source* or the *target?*

Originally 'truth' meant fidelity to the source. It was about loyalty to whomever empowers the truth-teller, be it the Christian deity or a Roman general. In this context, it was associated with executing a plan of action, be it in the cosmos or on the battlefield. One remained 'true' by following through on the power giver's intention, regardless of manner or outcome. It is this sense of 'true' that enabled the Jesuits, a Counter-Reformation Catholic order founded by a soldier, Ignatius Loyola, to do God's work by operating on the principle that 'the end justifies the means'.

However, thanks to another Catholic, Thomas Aquinas, truth came to be seen in the modern period as loyalty to the target – specifically, the empirical objects already in the field of play. His own Latin turn of phrase was *adequatio ad rem*, whose crude English translation, 'adequacy to the thing', captures the disempowering character of the concept, which philosophers continue to dignify as the 'correspondence theory of truth'. Aquinas, writing at a time of considerable heresy in the late thirteenth century, was reasserting confidence that the world as it normally appears is close enough to God's plan that the faithful should stop trying to second-guess God's intentions and focus instead on getting the empirical details of the Creation right. Today, Aquinas is the official philosopher of the church, a secure guide to the accommodation of science to faith.

These contrary pulls on the concept of truth – the source vis-à-vis the target – have persisted to this day. When Newton famously declared 'Hypotheses non fingo' ('I feign no hypotheses') in the second edition of *Principia Mathematica*, he was diverting suspicious religious readers who feared that he might be trying to get into 'The Mind of God' rather than simply providing a perspicuous account of nature's order. Of course, there is no doubt – given his voluminous private theological writings – that Newton was indeed aiming to second-guess the deity in which he believed. He was going for the source, not merely the target of all knowing. Against this backdrop, it is ironic that an avowedly 'atheist' physicist such as Stephen Hawking, successor to Newton's Cambridge mathematics chair, managed to parlay 'The Mind of God' as the driving metaphor of that popular science classic *A Brief History of Time*. Newton and Hawking differ not only in terms of details of execution but also degree of self-awareness. Newton was deliberately concealing what by Hawking's day had been formally disowned if not long forgotten.

One philosopher who offers guidance in navigating through the somewhat surreal post-truth intellectual environment is Hans Vaihinger (1852–1933), the person most responsible for turning Immanuel Kant into a fixture of scholarly interest, by founding *Kant Studien* in 1896. Vaihinger also developed an entire world view around Kant's repeated use of the phrase *als ob* (as if). Much of the normative force of Kant's philosophy comes from thinking or acting 'as if' certain things were true, even though you may never be able to prove them and they may even turn out to be false. Vaihinger (1924) called the resulting world view 'fictionalism', and it epitomizes the post-truth sensibility. And seen through Vaihinger's eyes, philosophy appears to be the most post-truth field of them all.

A good way to see Vaihinger's point is to consider contemporary philosophy's notorious schism between the 'analytic' and 'continental' schools. The analytics accuse the continentals of having picked up all of Friedrich Nietzsche's worst habits. The result is a trail of spurious reasoning, fake philologies, eccentric histories, obscurantism and hyperbole. This is quite a list of offences to the truth, yet it is striking that analytic philosophy's most lasting contributions have been a series of thought experiments, which are no more than figments of the imagination – such as Hilary Putnam's 'brains in a vat' or John Searle's 'Chinese room' – that are passed off as heroic abstractions from some hypothetical reality. The rest of analytic philosophy is basically just scholastic wrangling about the wording of these thought experiments and the conclusions one is licensed to draw from them, leavened by occasional moments of high dudgeon, as well as displays of ignorance, narrow-mindedness and bias vis-à-vis other, typically more 'continental' or 'postmodern', modes of reasoning.

Vaihinger could make sense of what is going on here. He divided our approach to the world into *fictions* and *hypotheses*. In a fiction, you do not know that you inhabit a false world, whereas in a hypothesis you know that you do not inhabit a false world. In either case, 'the true world' does not possess any determinate epistemic standing. On the contrary, you presume 'a false world' and argue from there. From this standpoint, continental philosophers are purveyors of fictions, and analytic philosophers of hypotheses. What we colloquially call 'reality' moves between these two poles, never really honing in on any robust sense of truth. Here one needs to think of 'fictions' on a sliding scale from novels to plays to laws ('legal fictions') and 'hypotheses' on a sliding scale from what Euclid was talking about to what scientists test in a lab to what people do when they plan for the future.

Does this mean that truth is a redundant concept altogether? That there is a 'redundancy theory of truth', proposed by the logician Frank Ramsey nearly a hundred years ago, suggests as much. Moreover, the theories of truth that have followed in its wake – alternatively called 'deflationary', 'disquotational', 'expressive' and even 'honorific' (to recall Richard Rorty's reappropriation of John Dewey) – can be added to the post-truth repertoire of analytic philosophy (cf. Haack 1978: chap. 7). But 'in fact' (permitting the locution), Vaihinger would say – and I agree – that truth turns out to be whatever is decided by the empowered judge in the case at hand. In other words, Bacon was right, after all, which perhaps explains why Kant dedicated the *Critique of Pure Reason* to him. But what sort of 'power' are we talking about here? In the next section, which centres on the philosophical origin of the post-truth imaginary, I focus on the idea of *modal power*, which involves control over what people take to be possible. When Otto von Bismarck said that politics is the art of the possible, he was acknowledging Plato's discovery that all power boils down to modal power.

The Birth of Rhetoric as the Crucible of the Post-Truth Imaginary

The difference between philosophy and rhetoric as academic fields of study has been increasingly institutionalized in the modern period. Actually, this is to put the point too weakly. It would be better to say that the fates of philosophy and rhetoric in the academy have been inversely related during this period. Philosophy has become a formidably serious discipline dedicated to 'The Truth', while rhetoric has come to be seen at best as a handmaiden of The Truth, sometimes appearing as merely ornamental while at other times outright subversive. In many respects, this is exactly as Plato would have wanted it. Moreover, there is a cottage industry of scholars who puzzle over how Plato – through his mouthpiece Socrates – managed to pull this off against the Sophists, the local post-truth merchants in fourth-century BCE Athens. The Sophists would have rather never seen the distinction between philosophy and rhetoric be drawn. It should be clear that I am on their side of the argument.

Contrary to the shared assumption of the scholars who have pondered these matters, I believe that Plato and the Sophists differed less over what counts as good 'dialectical' (i.e. philosophical or rhetorical) practice than over whether its access should be free or restricted. Indeed, I would have us see both Plato and the Sophists as post-truth merchants, concerned more with the mix of chance and skill in the construction of truth than with the truth as such. Focusing on Plato's hostility to playwrights, I argue that at stake is control over 'modal power', which is ultimately defining the sphere of what is possible in society. I end with a brief discussion of the problematic of public relations as an ongoing contemporary version of much the same story.

About a quarter-century ago, Edward Schiappa and John Poulakos had a rather peculiar debate about the origin of rhetoric that was spread over several books and journals (e.g. Schiappa 1990a; Schiappa 1990b; Poulakos 1990). What made the debate peculiar was that it seemed to turn on a pedantic point that Schiappa prosecuted with great thoroughness, namely, that 'rhetoric' as a noun did not exist prior to Plato's coinage in the *Gorgias*. But more was clearly at stake, as Poulakos fully realized: the potential denial of a self-consciously understood rhetorical tradition that is not subordinate to the philosophical tradition. This explicitly counter-Platonic sense of rhetoric had long been identified with the 'Sophists', teachers of the verbal arts who figure prominently as foils in the Socratic dialogues. Yet, even defenders of Sophistic honour have had to admit that the very idea that these teachers shared a common world view largely comes from their finding a common nemesis in Socrates. In effect, Schiappa accused Poulakos and other pro-Sophists of falling for Plato's own rhetoric by treating a rather disparate group of characters as if they constituted a coherent school of thought. Moreover, as if to add insult to injury, Schiappa also seemed to suggest that close attention to Plato's original usage reveals that it largely anticipated analytic philosophy's disparagement of 'rhetoric' as unnecessary, deceptive and fallacious.

In the context of this debate, I am inclined to agree with Poulakos's verdict that Schiappa's focus on the letter of the Greek blinds him to the spirit of the Greeks. But I am less sure about Poulakos's own positive claims for an autonomous Sophistic tradition. After all, Plato's Socrates and his Sophistic antagonists are not so very different in the sort of command they have over the *logos*. Indeed, this is arguably why they so readily take each other as dialogic equals – which differs markedly from the epistemic authority that they exert over their respective fans and clients. However, what really bothered Plato about the Sophists as a group – beyond whatever doctrines they might or might not have shared – was that they sold their dialectical skills to whomever was willing to pay, regardless of genuine need. I have previously suggested that the modus operandi of the Sophists may have anticipated 'marketing' in the modern sense of supplying goods that perpetuate their own demand, as the biggest selling point of the Sophists' rhetorical services was that their possessor would be provided with a competitive advantage, which by definition was temporary and hence would always need to be topped up (cf. 'lifelong learning'). This was quite at odds with the view of markets found in Aristotle, which was as occasional clearing houses for the mutual cancellation of household surpluses and deficits – but in any case, markets were not ends in themselves (Fuller 2005b: chap. 1).

Diagnosed in more familiar rhetorical terms, the Sophists' great sin was their indifference to the character of the potential client. To be sure, Socrates never begrudges the Sophists' wanting to make a living. He even grants that what they have to teach is somewhat valuable. However, Socrates does chastise the Sophists for overselling, given the sort of people who are likely to be attracted to their services. In this respect, Socrates is the original enemy of the free market. He believes that only people with the right character can be taught the sort of skills that the Sophists can reasonably teach. Indeed, in correspondence with Leo Strauss, Alexandre Kojève suggested that Plato was promoting a kind of metalevel clearing house in which those who know how to rule transact with those with the power to rule, each serving the other's legitimate needs, resulting in a whole greater than the sum of its parts (Strauss 2000). The implied conclusion, on which Plato based his own career, is that the relevant market is necessarily small. Moreover, by the time Plato wrote his works, the destabilizing effects of widely distributed Sophistic teachings had already taken their toll on Athens's domestic security. But significantly, Plato's response was not to eliminate the Sophistic teachings entirely but to restrict access to them to a specially prepared class of future leaders, the 'philosopher-kings', as described in the *Republic*.

This strategy helps explain Plato's preoccupation with censorship, and especially his proposed ban on 'poets', though he might as well have said 'Sophists' – which is to say, anyone with skills that might compete with those needed to succeed as a philosopher-king. In that sense, it might be said that Plato coined the term 'rhetoric' less because he wanted to distinguish good from bad philosophy per se and more because he wished to concentrate the full skill set of philosophy into one source of power and control. We might call this, by analogy with 'monopoly capitalism', *monopoly intellectualism*. Thus, when Socrates famously dismisses the art of writing in the *Phaedrus*, he should be understood as objecting at least as much – if not more – to the *individual* use of writing as to writing per se. After all, writing holds the potential to render each person a law unto him- or herself, once each person is able to compose scripts designed to direct his or her speech and action. While Socrates himself stressed the resulting inauthenticity of expression, the prospect of a society of literate individuals made the governing task for any aspiring philosopher-king that much more difficult, as the people's expressed words and deeds would be subject to a level of mediation that would render their thoughts less transparent – and hence less manipulable – than they otherwise might be.

To take another example, one key rhetorical strategy deployed by Plato in 'The Allegory of the Cave' in Book 7 of the *Republic* involves the shifting of perspective between what the people of the light and the people of the dark see, which is presented with an eye to enabling the former to draw out the latter. This ability to reverse perspective was a dialectical skill the Sophists commonly taught. However, because the skill was made available to all paying customers, the net effect of possessing the skill was that people could second-guess each other's moves, which conferred on politics a game-like feel, in which a rhetor endlessly pivots between pre-empting and adapting to the interlocutor, who is invariably cast as an opponent. Under such volatile, even febrile, conditions, there is little opportunity or incentive for the dialectician to realize anything resembling a Hegel-style dialectical synthesis, or *Aufheben*, whereby opponents find a space to set aside their differences to achieve something greater than they could achieve either left to their own devices or in competition for the same position (Melzer 2014: chap. 3).

I have so far managed to discuss Plato and the Sophists without mentioning the word 'truth'. That is because both were, in today's terms, 'post-truth' in orientation. They were less concerned with the truth itself than with the conditions that make truth possible. An intuitive way to see this point – and to see how Plato and the Sophists differed over it – is to observe that both the Sophists and Plato saw politics as a game, which is to say, a field of play involving some measure of both chance and skill. However, the Sophists saw politics primarily as a *game of chance*, whereas Plato saw it as a *game of skill*. Thus, the sophistically trained client deploys skill in aid of maximizing chance occurrences, which may then be converted into opportunities, while the philosopher-king uses much the same skills to minimize or counteract the workings of chance. In what follows I focus mainly on the *playwright* as a particular sort of 'poet' who challenges the Platonic enterprise at its heart by explicitly taking on this distinction.

The performing arts are the natural vehicle for the post-truth world. The classic contrast between Plato and Aristotle on the theatre's normative standing makes the point. They were basically two sides of the same coin. Plato was deeply suspicious of actors and playwrights because of their capacity to blur the audience's sense of the difference between the actual and the possible, whereas Aristotle welcomed dramatic performances – but only if a play's plot was fully resolved by the time the action onstage was concluded. Thus, neither welcomed any spillover effects from the theatre to the larger society. However, Plato considered them an outright political threat, whereas Aristotle saw them simply as an aesthetic failure. This difference perhaps reflected the fact that Plato was writing while Athens was still an independent democracy, whereas Aristotle was writing after the city had fallen to his employer, Alexander the Great. In other words, there may have been much less to play for by the time Aristotle wrote about the theatre's normative standing.

Why is blurring the difference between the *actual* and the *possible* – or in the somewhat different way we would now cast it, *fact* and *fiction* – so politically threatening? Let us start by recalling the two general types of grounding for moral theories: on the one hand, by appeal to a lawmaker who is either human or specifically oriented to humans; on the other, by appeal to 'nature', understood as an entity that is either indifferent to humanity or interested in humans only as part of a larger order. The tradition that includes Moses, Jesus, Kant and Jeremy Bentham belongs to the former category. The tradition that includes Aristotle, Epicurus, David Hume and today's evolutionary psychologists belongs to the latter. As for Plato, he notoriously held a divided judgement on this matter. Yes, he believed in the truth of the former grounding, and offered to train a class of potential political leaders – the philosopher-kings – in this understanding. But equally Plato believed that the smooth functioning of the state required that everyone else believed in a 'natural' grounding for morals.

In the histories of theology and philosophy, Plato's Janus-faced position is generally called *the double-truth doctrine* (Melzer 2014: chap. 2). It marks him as a post-truth thinker – not because he denies the existence of truth. Not only does Plato not deny the existence of truth but he also pledges personal allegiance to it. Rather Plato's post-truth sensibility is revealed by the circumscribed role he assigns to truth in human affairs, a scare-quote sense of 'truth' that enforces the epistemic superiority of the belief system of the elites over that of the masses, which in his view allows for more effective governance. In effect, the double-truth doctrine prescribes that it is in the interest of both the elites and the masses that there be two 'truths', the one that ranges over every possible world in which we might live (i.e. the post-truth sense of 'truth' of the governors) and the other that reassures everyone that the world in which we currently live is a just one.

Until the Protestant Reformation initiated the spread of literacy in Europe, the Platonic sense of elite truth was restricted to those few who could write and read. They could seriously aspire to inhabit the mind of God, which provided the gold standard for how to understand conjuring with possible worlds. After all, the very idea of possible worlds is rooted in the capacity to interpret a text in multiple ways, which in turn requires literacy. But with the onset of mass literacy, the Platonic sensibility began to be democratized. The 'post-truth' horizon of the Platonic elites came to be 'privatized', as more people were able to read a text for themselves, interpret it accordingly and legislate for themselves. What came to known as 'public' was the truth prescribed on the field, the rules of the game in play that were collectively ratified and upheld by this plurality of Platonists. It was the stuff of 'social contracts' and 'constitutions'. As this phenomenon came to be noticed in the eighteenth-century Enlightenment, 'hypocrisy' figured as an ambivalent term for those who publicly played by the rules but tried to game it to their private advantage (Sennett 1977). Parliamentary and congressional politics have continued this democratized Platonic attitude to this day. However, we are now living in a period of further democratization, associated with the rise of cyberliteracy, the so-called 'post-truth condition'.

But let us rewind back to Plato. From his standpoint, performing artists stand guilty of providing a vivid sense of any number of 'unnatural' orders, which is to say, alternative worlds that could inspire the audience to enact them through sheer force of personality and will once they leave the theatre. Of course, that is also precisely how the lawmaker brings about his 'natural' order – but he is not normally facing competition, which could force him to reveal his hand. Plato, followed by Machiavelli, believed that the security of political fabric woven by the lawmaker is best maintained by the seams of its construction remaining hidden – as opposed to being revealed as no more than an improvisation from a script that could read differently to produce an alternative version of reality (cf. Goodman 1978; Fuller 2009: chap. 4).

As might be expected from the man who revolutionized philosophy by adopting the dramatic form, Plato understood the playwrights perfectly – and they him. They are each other's nemesis in a post-truth world, where 'the true believers', so to speak, are the rest of society, the 'masses' who regard the law-like character of their world as the reflection of some natural order— that is to say, *not* as the resolution of opposing forces in a power struggle. Specifically, Plato and the playwrights are struggling over *modal power*. I allude here to the branch of metaphysics called 'modality', which in the modern period has been often understood as a kind of second-order logic.

Put in most general terms, modality is about *how* something exists in the world: if some proposition *p* is true, just how true is it? Even if we agree that *p* is indeed true, it makes a big difference whether we believe that *p* is *necessarily* true or only *contingently* true. Of course, most of what we believe is believed as contingently true, but contingency itself offers a spectrum of possibilities, ranging from the virtually necessary (i.e. it is very difficult to conceive of a condition in which the truth of *p* would be overturned) to the virtually false (i.e. it is very easy to conceive of a condition in which the truth of *p* would be overturned). The Platonic political order is maintained by the masses believing that the contingent is virtually necessary. In contrast, theatre is the crucible for converting the contingent truths of the lawmaker – and the status quo more generally – into the virtually false by the onstage construction of plausible alternative ('virtually true') realities to the ones normally on display outside of the theatre.

This understanding of modal power puts Plato's famed discussions about the nature of justice in a somewhat different light, especially when expressed by his uncensored mouthpiece, Socrates. A touchstone is a quote fabricated by Thucydides in book 5 of the *Peloponnesian War*. An Athenian diplomat tells his soon-to-be vanquished counterpart that 'right, as the world goes, is only in question between equals in power, while the strong do what they can and the weak suffer what they must'. This quote is seen as the precedent for Socrates's dialectical encounter in the *Republic* with Thrasymachus, who argues that 'might makes right'. The tenor of Socrates's interrogation is not quite as it is popularly presented. Socrates takes Thrasymachus to be claiming that the right to rule is implied in the very idea of being mighty, in response to which Socrates says simply that the right to rule goes to the individual who finds that no one else would make a better leader – and the ruler's might then derives from *that* fact. In short, Socrates turns Thrasymachus's claim on its head: *right makes might*.

But exactly how should we analyse the Socratic counterclaim? After all, 'right' in this context seems to mean little more than 'comparative advantage'. Moreover, it would be anachronistic to suppose that Socrates was alluding to some established democratic procedure that confers the right to rule on the basis of, say, an electoral outcome. The Athenians were all for voice votes and selection by lot. Rather, he means something much looser and game-like. 'Might' is measured in terms of the perception that serious alternatives are lacking, which is to say, that the candidate is 'virtually necessary' and on that basis is entitled to rule. How that virtual necessity is publicly demonstrated is an open question, but Socrates can be reasonably read as saying that it involves the consent of the ruler and ruled. In other words, the two sides come to agree by some means or other – peaceful or violent, discursive or intuitive – that one side is better suited to rule, which is to say, that the interests of the two sides are optimized by the agreement.

The natural descendant of this sensibility in our both more democratic and more economistic times is John Rawls's (1971) 'difference principle', which lies at the foundation of his famous theory of justice. It says that inequality is justified only if the worst off benefit more than if the inequality were not in place. But this presumes that the people can agree – by whatever means – on judgements of this sort, including when it comes to recognizing the legiti-macy of subordinating themselves to a 'superior'. The Platonist's fear is that playwrights may well destroy the delicate sense of 'virtual necessity' that lies behind such judgements by persuading the audience that what is true today could be false tomorrow, and vice versa: thus, the 'necessity' is overshadowed by the 'virtual'. What playwrights threaten to reveal – and what Plato would rather see hidden – is the repression of the imagination that is required for the regime of the lawgiver to correspond to the people's sense of a natural order.

Consider that from the strictly dramaturgical standpoint, the highest com-pliment that can be paid to a play is that the action onstage is so 'realistic' that one could easily imagine it happening offstage. But it is precisely at that point that the fact/fiction distinction has been blurred in a dangerous way for Plato's 'law and order' world. In terms of modal power, the fact/fiction distinction is strongest when it is easy to assign degrees of uncertainty to states of the world along a spectrum from necessary (the modal proxy for 'certainly true') to impossible (the modal proxy to 'certainly false'). This is ordinarily experienced as a kind of common sense about courses of action that are and are not 'risky' to undertake. However, when everything seems equally (un)certain, then the fact/fiction distinction has become truly blurred and the playwright has tri-umphed over the philosopher-king. Thus, people will have come to undervalue the risk entailed by a fundamental change of course and hence be more open to it, if not outright initiate it, and thereby threaten the philosopher-king's authority – and, so Plato himself thought, the social order itself.

Many accounts of Greek etymology observe that *theory* and *theatre* share a common ancestry in *theos*, the Greek word for God. Implied here is a concep-tion of the deity whose supremacy rests on being able to see both inside and outside its own frame of reference, a double spectator, or *theoros*. In logic, this is called *second-order* awareness: one not only plays a language game but also knows that the game is only one of many that she might be playing. In what follows, I associate this awareness with the *post-truth* mentality. In Plato's dia-logues, the sophists are clearly trying to cultivate just such a mentality in their clients, which in principle would give them a god-like discretion to decide which game they play in the open space of the agora. Socrates pushes back from this arch sense of self-awareness, arguing that there is ultimately only one game in town, *truth*, adherence to which would keep everyone playing by the same rules, thereby sticking to what logicians would call a *first-order* awareness.

Perhaps the most straightforward example of Socrates's approach appears in chapter 20 of *Protagoras*, in which he manages to get his sophistic opponent to admit that all virtues are one because they all have the same contrary, *aphrosyne*, which is normally translated as 'lack of proportion or perspective'. In the process of persuading Protagoras of this thesis, Socrates gradually removes the sense of the virtues as something skill-like, each possessing its own gradient along which one may perform better or worse. This serves to neutralize the image of the citizen that the sophists presupposed, namely, one whose competence consists in playing the virtues off each other as the situation demands, very much in the spirit of a modern economic 'optimizer' who decides to act after having traded off her various interests. In its place, Socrates proposes that to think that there are separate virtues is to reflect one's ignorance of what virtue is. Thus, the just, the good, the beautiful and so on are all simply aspects of virtue as such. 'Virtue' in this univocal sense is identified with the truth, in that everything is understood in its rightful place. From this standpoint, by proliferating virtues as skills, Protagoras is selling parts as if they were the whole. Socrates's argument, as filtered through Plotinus, would exert profound influence in the Middle Ages, as the Abrahamic deity came to stand for what Socrates had identified as the one truth behind all its virtuous appearances.

The ultimate difference between Socrates and the sophists is not the dialectical capacities of the two sides, which are basically the same. In this respect, Plato's coinage of 'philosophy' for Socrates's argumentative style and 'rhetoric' for the sophistic style is itself a rhetorical diversion. Rather, the difference lay in Socrates's objection to the free-wheeling – some might say 'democratic', others might say 'commercial' – way in which the sophists deployed these common capacities. To render the premises of arguments pure inventions of the arguer is potentially to turn any human into a deity if enough people are persuaded to regard his or her premises as the rules of the game by which they all subsequently play. And as Plato knew from first-hand experience, the sophists did succeed in persuading enough citizens of their 'dialectical divinity', so to speak, that Athenian democracy ended up reproducing the chaotic sociability of the gods of Greek mythology. Unfortunately, in the real world this led to the defeat of Athens at the hands of Sparta in the Peloponnesian War, the beginning of Athens's terminal decline.

In Plato's telling, Socrates stands for the need to play just one game, which explains why the sophists and the 'poets' are put in the same basket. I put 'poets' in scare quotes because the term should be understood in its original Greek sense of *poiesis*, the productive use of words to conjure up worlds. In

Plato's day, playwrights were the poets of chief concern, but in our own day those adept at computer coding – 'hackers' – might be the main source of comparable subversion (cf. Wark 2004). Whereas Plato believed that only philosopher-kings in training should cultivate a second-order imagination – think Hermann Hesse's *The Glass Bead Game* – his enemies were keen to distribute this capacity as widely as possible. At stake was *modal power*. In other words, whatever the rules of the game happen to be – or seem to be – they could be otherwise. From the post-truth standpoint, truth looks like an extreme disciplining of the imagination, which accounts for the authoritarian – and even totalitarian – feel of Plato's positive proposals for governing the polity in the *Republic*.

At the same time, it is easy to see how Augustine and other early Christian thinkers found Plato attractive as a metaphysical backdrop for their own monotheistic views, since the dialogues brought into sharp relief the perils of humans trying to behave like gods in a polytheistic cosmos. However, Christianity is also a religion that claims that believers have some sort of direct contact with their one deity. Indeed, since humans are biblically created *in imago dei*, a phrase Augustine himself popularized, the prospect that each person might not merely imitate a god – as in the Greek case – but actually instantiate divinity proved to be an endless source of heresies. It culminated in the Protestant Reformation, which arguably has reproduced the sophistic situation that the early Christians were trying to avoid by embracing Plato. Of course, Augustine's own original solution was to strengthen the authority of the established 'universal' (*katholikos* in Greek) church of his day, which over roughly the next thousand years held together notwithstanding two major schisms across the European continent, the first East-West (resulting in the Orthodox churches) at around 1000 AD and the other North-South (resulting in the Protestant churches) at around 1500 AD.

The Modal Power of the Entertainment Industry

In this context, it is worth recalling that 'entertainment' is an early seventeenth-century English coinage designed to capture an abstract sense of tenancy, as in the case of the king who keeps a poet or playwright on retainer for purposes of amusement but whose proximity ends up exerting influence over his political judgement. Plato would have recoiled from such an innovation – as did Thomas Hobbes, who at the time suspected the theatricality of experimental demonstrations, a singular mode of royal entertainment, the ascendancy of which form the basis of the most influential monograph in the historical sociology of science in recent times (Shapin and Schaffer 1985; cf. Shaplin 2009, a dramatization of Hobbes's concern). The fear evoked – or opportunity afforded – by entertainment is that after the final curtain is drawn, the audience might themselves continue acting in the spirit of the performance they had observed, effectively turning 'real life' into an extension of the stage or, as Hobbes feared, the lab.

With regard to science in our own day, Hobbes still has plenty – if not more – to worry about. Researchers still do not take proper notice of the role that the capacity to stage, re-enact or simulate events in digital media is acquiring as part of the scientific validation procedure, alongside the controlled laboratory experiment and more conventional computer modelling techniques. Indeed, *Scientific American* editor John Horgan (1996) notoriously argued that as the site of scientific research has moved from the field to the lab to the computer, its modus operandi for evaluating knowledge claims has increasingly come to resemble literary criticism, in terms of privileging aesthetically oriented mathematical criteria.

In what David Kirby (2011) has aptly called 'Hollywood knowledge', consultants shuttling between the scientific and cinematic communities do not merely convey information that each needs to know of the other to get their respective points across. Rather, the scientists and the film-makers mutually calibrate their goals and standards of achievement. In particular, film-makers not only shape the expectations of their viewers but also fuel the scientific imagination itself, as aspects of complex concepts and situations are heightened, sometimes exaggerated, but in any case extended to their logical extreme – very much in the spirit of the best thought experiments. This development deserves further serious epistemological exploration, as it breaks down a distinction much cherished by sociologists of expertise (e.g. Collins and Evans 2007), namely, between those whose expertise is merely 'interactional' (i.e. the sort of critical insight that an amateur might offer a professional based on, say, familiarity with the relevant literature) and 'contributory' (i.e. the capacity to make an original contribution to the field).

A striking example is Fritz Lang's 1929 proto-feminist film, *Frau im Mond* (Woman on the moon), many aspects of which – from the design of the launch pad to the 'countdown' ritual – were adopted by NASA 30 years later, partly under the influence of Wernher von Braun, who had seen the film as a youth. More generally, the changing image of DNA – from, say, the simple combinatorial depiction of the molecule on a primitive computer screen in the 1973 BBC television series *The Ascent of Man*, to the three-dimensional dynamic machinery one routinely sees today in science films – illustrates how improvements in media representation have helped reorient both scientific and lay understanding of what fundamental concepts are supposed to be about. In this respect, to adapt the old Gil Scott-Heron song, the next scientific revolution may well be televised.

The very idea that scientific theories might stand or fall by whether they make for, say, a good cinematic experience may sound frivolous at first, but is it so very different from judging the soundness of mathematical models by the elegance of the simulations they produce on a computer screen? In thinking about such proposed seismic shifts in scientific representation, it is always helpful to take the long view. Recall that among the early objections to the use of experiments as the means for resolving scientific disputes was that they involved too much behind-the-scenes staging and editing, which resulted in an artefact with no clear correlate in nature that simply served to manipulate the senses. Indeed, in many cases, the experiments could not be strictly replicated but instead counted on the impressed observer coming to see nature through the principles or phenomena allegedly demonstrated by the experiment. Objections of this sort – notoriously lodged by Hobbes against Robert Boyle – are usefully seen as an early modern version of Plato's ancient critique of poetry and drama as fraudulent imitation of authentic knowledge and sentiment. So too we may come to see the current scepticism surrounding 'science-by-television', once people with the appropriate media production skills join the ranks of scientific validators, following in the footsteps of artisans, mechanics and programmers.

My post-truth sense of entertainment's intellectually empowering character goes against the grain of Neil Postman's (1985) influential demonization of its alleged narcotic effects. To be sure, Postman was fixated on television, which he understood as Marshall McLuhan's absorbing yet non-interactive 'cool' medium that effectively sucked the life out of its viewers, a process that had been recently sensationalized by David Cronenberg in the 1983 film *Videodrome*. But rather than the vampire, Postman might have considered the *virus* as the model of entertainment's modus operandi, whereby the host is not so much annihilated as simply contaminated by the guest organism. This then gets us back to the problem that originally concerned Plato, one which Antoine Artaud's (1958) 'theatre of cruelty' converted into a virtue: it is not that the poets send the audience into a dream state but that the audience might enact those dreams in 'real life'. The normative limits of 'reality television' provide an interesting contemporary benchmark on this issue. Whereas television producers and audiences are enthusiastic about *Dragons' Den*-styled programmes (called *Shark Tank* in the United States) that cast entrepreneurship as a talent competition, similarly styled proposals to stage political elections have been met with the sort of disapproval that would have caused Plato to breathe a sigh of relief – for the time being at least (e.g. Firth 2009).

In the end, the philosopher-king is worried less that the people have seen through his fabrications concerning the natural basis of the social order than whether they are prepared to cope with the consequences of acting in a world in which the workings of the law and nature are no longer seen as so closely intertwined, let alone mutually reinforcing. In short, the post-truth problem is not the truth itself but people's ability to handle it. After all, assuming truth exists, the flip side of saying that the truth is independent of human conception is saying that humans may not be up to the task of facing the truth. Not surprisingly perhaps, Henrik Ibsen, who is widely recognized as the greatest dramaturgist of modern times, centred many of his plays on a 'life-lie' that is uncovered in the course of the plot, which results in some literal or metaphorical violence. Ibsen's great talent lay in orchestrating his characters in such a way that they unwittingly elaborate a life-lie over the course of the play, all the while adumbrating that they will eventually come to see it as what it is. As Ibsen's early admirers – notably George Bernard Shaw – realized, this was a relatively polite way to destabilize the existing order, a way that would earn even the grudging respect of Plato, who after all was the original master of dialogue. (It also explains Shaw's often heavy-handed Socratic efforts in his own plays, baldly asserting what Ibsen could have more subtly expressed.)

Whatever success the Platonists have had over the playwrights in the political arena, their grip has slipped during the history of capitalism, especially as sheer financial speculation played an increasing role in the circulation of capital. Financial speculation requires a reconstruction of the economy in the image of the playwright, who ultimately wants people to see that the future need not be like the past. Thus, well-established companies may well buckle from the competitive pressure of ambitious upstarts, if the latter are given sufficient support by investors who become caught up in the same aspirations. Platonists – not least John Maynard Keynes – tend to see this in terms of 'animal spirits' elicited from the masses, who are quick to respond to whatever promises to harm or benefit their material condition. However, such investors may be seen in quite the opposite light, namely as exercising their rational will over an inherently turbulent world to bring about a better state of affairs. From that standpoint, the status quo may well have passed its sell-by date, and it could be time to consider a different sort of future. Whatever the interpretation, investors trade off the security of the present for the prospect of greater freedom (in the form of greater disposable income from higher returns) by shifting their assets. Of course, whether this strategy works depends on whether the upstart companies perform as they promise; otherwise, the investors will have been made poorer in the process. But in either case, the Sophists would claim a second-order victory, as the economy effectively morphs from being a game of skill to a game of chance.

To conclude, it is worth remarking on the person who made the most impressive Platonic response to the rise of finance capitalism by focusing specifically on the key role that advertising played in promoting products that had yet to prove themselves in the market. It came from the great twentieth-century theorist of 'objective journalism', Walter Lippmann (1922). He began his career working with a contemporary who became the father of modern public relations, Edward Bernays (1923), in Woodrow Wilson's successful campaign to persuade Americans to fight in the First World War. This was itself quite a remarkable public relations feat, considering that the United States was never officially attacked and that the war itself was happening more than three thousand miles away on a continent, Europe, that this nation of immigrants had been more than happy to leave.

Both Lippmann (1925) and Bernays (1928) appreciated the significance of their rhetorical coup, and their parallel fates over the rest of the twentieth century have been brilliantly captured by the BBC documentary film-maker Adam Curtis in *The Century of the Self*. In a nutshell, Bernays took advantage of America's laissez-faire ethos in the roaring twenties to turn the successful war campaign strategy into a billion-dollar peacetime industry, whereas Lippmann called for state licensing to inhibit advertisers from enticing consumers to purchase goods, services and – last but not least – stocks that end up only failing to deliver on their promises. In one of the earliest reported cases of 'spin', Bernays managed to flip Lippmann's pejorative description of public relations as the 'manufacturing of consent' into the more professional-sounding 'engineering of consent' (Jansen 2013). Of course, Noam Chomsky much later succeeded in restoring Lippmann's original meaning (Herman and Chomsky 1988).

Nevertheless, Bernays had the upper hand in a rapidly upscaling broadcast media environment of the 1920s, even though Lippmann's original concerns were vindicated as early as 1929 by the Wall Street stock market crash, which led to the Great Depression across the capitalist world. Nevertheless, the Platonists have remained on the back foot, as evidenced in finance capitalism's subsequent boom and bust cycles, often fuelled by the ultimately unfulfilled promises of innovation. To be sure, the state regulators eventually make their appearance, but only after the fact and ruing the Platonists they failed to heed who Cassandra-like had predicted the latest catastrophe. While this is not the image of Plato that we are used to seeing, it is what Plato looks like to a post-truth world that has made his demonized sense of 'rhetoric' its own.

How Truth Looks to Post-Truth: Veritism as 'Fake Philosophy'

It is worth stressing that a 'post-truther' does not deny the existence of facts, let alone 'objective facts'. She simply wishes to dispel the mystery in which the creation and maintenance of facts tend to be shrouded. For example, epistemologists have long tried to make sense of the idea that 'correspondence to reality' explains what makes a particular statement a 'fact'. On the most ordinary reading, this sounds a bit mysterious, since it suggests a strange turn of events. Take the case of scientific facts: (1) Scientists do whatever they do in a lab. (2) They publish something that convinces their learned colleagues that something happened there, which sets off a train of actions that starts by imprinting itself on the collective body of scientific knowledge and ultimately on the world at large as an 'expert' judgement. (3) Yet – so the 'truthers' tell us – in the end what confers legitimacy on the fact (i.e. makes it 'true') is something *outside* this process, a reality to which it 'corresponds'.

To someone not schooled to 'know better' (i.e. in the 'truth' mode), (3) seems to be quite an arbitrary conclusion to reach, just given (1) and (2). Unsurprisingly then, the twentieth century has been largely a story of philosophers gradually falling out of love with this scenario, which in turn has animated the post-truth sensibility. Indeed, there is a fairly direct line of intellectual descent from the logical positivists and the Popperians to contemporary social constructivism in the sociology of scientific knowledge, contrary to their textbook representation as mutual antagonists. I have gone so far as to call science and technology studies (STS) 'postmodern positivism' – in a non-pejorative sense (Fuller 2006b)! To put the reader in the mood for what follows, consider the career of Ludwig Wittgenstein (1889–1951), whose two-phase career runs like a red thread through all these movements.

The two phases of Wittgenstein's philosophical career are fairly described as attempts to define truth from a 'truth' and a 'post-truth' perspective: the early phase epitomized by the *Tractatus Logico-Philosophicus* and the later phase by the *Philosophical Investigations*. The early writings were fixated on the idea of a truth-functional logic into which all meaningful statements might be translated and evaluated. If a statement was deemed 'meaningful', then one could determine straightforwardly – perhaps even mechanically – whether it was true or false. In contrast, the later writings were concerned with the fact that the same string of data, be they quantitative or qualitative, may be subsumed or interpreted under any number of rules that would render them meaningful. In that case, the form of inference required is closer to abduction than deduction.

The early Wittgenstein captures the 'truth' orientation, whereby the rules of the knowledge game are sufficiently well understood by all players that appeals to 'evidence' mean the same thing to everyone concerned. This is the world of Kuhnian paradigms, whose knowledge game is called 'normal science' (Kuhn 1970). To be sure, depending on the state of play some evidence may count more than others and may even overturn previous evidence. However, the uniformity of epistemic standards means that everyone recognizes these moves in the same way and hence there is a common understanding of one's position in the epistemic tournament, including which teams have made the most progress.

The later Wittgenstein captures the 'post-truth' orientation, whereby the knowledge game is not determined by the rules; rather, determining the rules is what the knowledge game is about. Emblematic of this approach is the duck-rabbit gestalt that appears not only in this period of Wittgenstein's work but also in Kuhn's account of the psychology of the 'paradigm shift' that characterizes a scientific revolution. The idea is that the same evidence can be weighted differently depending on the frame of reference adopted, which may result in a radical shift in world view. Both Wittgenstein and Kuhn agreed that whichever frame prevails is not preordained but a contingent matter, one which tends to be covered up after the fact by the justificatory narrative that the community concerned tells itself in order to go forward collectively. Kuhn notoriously dubbed this narrative 'Orwellian', after the work done in *1984*'s Ministry of Information, whereby a regular rewriting of history to match the current direction of policy subtly erases any memory that policies had been otherwise – and might also be otherwise in the future.

Where the later Wittgenstein and Kuhn differed was that the former appeared to think that the rules of the game might change at a moment's notice, depending on who is in the room when a binding decision needs to be taken. Thus, in principle the number series that begins 2, 4,... may continue with 6, 8 or 16, depending on whether the implied rule is agreed to be $n+2$, $n2$ or n^2. Usually there is precedent for settling the matter, but that precedent amounts to no more than a 'convention'. The alternative rules for going forward on such an occasion are comparable to the alternative dialectical framings of a situation that the sophist juggles at any given time until an opportunity presents itself (*Kairos*). Indeed, this interpretation served to make the later Wittgenstein the darling of ethnomethodologists in the 1970s and '80s, including within the nascent 'sociology of scientific knowledge', who launched STS.

In contrast, Kuhn believed that the decisive moments require a specific prehistory, the logic of which effectively forces a decision that is then taken only with great reluctance and may involve a de facto rejection of those who had been previously part of the relevant community. This 'logic' is characterized by the accumulation of unsolved puzzles in the conduct of normal science, which then precipitates a 'crisis', resulting in the paradigm shift that imposes new rules on the science game. In this respect, Kuhn might be seen to strike a balance between the two Wittgensteins or – to recall our earlier discussion – Socrates and the sophists.

The telltale sign is that they all define 'truth' as something *inside* – not outside – the terms of a language game. Put another way, 'truth' shifts from being a substantive to a procedural notion. Specifically, for them 'truth' is a second-order concept that lacks any determinate meaning except relative to the language in terms of which knowledge claims can be expressed. (This is known as the 'Tarski convention of truth'.) It was in this spirit that Rudolf Carnap thought that Kuhn's 'paradigm' had put pragmatic flesh on the positivists' logical bones (Reisch 1991; cf. Fuller 2000: chap. 6). (It is worth emphasizing that Carnap passed this judgement before Kuhn's fans turned him into the torchbearer for 'post-positivist' philosophy of science.) At the same time, this orientation led the positivists to promote – and try to construct – a universal language of science into which all knowledge claims could be translated and evaluated.

More to the point, the positivists did not *presuppose* the existence of some univocal understanding of truth that all sincere enquirers will ultimately reach. Rather, truth is just a general property of the language that one decides to use – or the game one decides to play. In that case 'truth' corresponds to satisfying 'truth conditions' as specified by the rules of a given language, just as 'goal' corresponds to satisfying the rules of play in a given game.

To be sure, the positivists complicated matters because they also took seriously that science aspires to command universal assent for its knowledge claims, in which case science's language needs to be set up in a way that enables everyone to transact their knowledge claims inside it, hence, the need to 'reduce' such claims to their calculable and measurable components. This effectively put the positivists in partial opposition to all the existing sciences of their day, each with its own parochial framework governed by the rules of its distinctive language game. The need to overcome this tendency explains the project of an 'International Encyclopedia of Unified Science'. In this respect, logical positivism's objective was to design an epistemic game – called 'Science' – that anyone could play and potentially win.

Perhaps the most elaborate 'fake philosophy', so to speak, that has been designed to counteract the post-truth sensibility is called *veritism*, which reasserts the 'outside' conception of truth by declaring it a necessary constraint if not the primary goal of any legitimate enquiry. Veritism is popular, if not dominant, among theorists of knowledge and science in contemporary analytic philosophy, most notably Alvin Goldman (1999). The extramural admirers of this doctrine include those keen on shoring up the epistemic authority of 'scientific consensus' in the face of an increasing multitude of dissenters and sceptics. In what follows below and the next chapter, I single out Erik Baker and Naomi Oreskes (2017) for special treatment because the simplicity of their formulations results in a clarity not normally observed by professional – yes, analytic – philosophers. The 'fakeness' of veritism comes from its studied refusal to engage with the essentially contested nature of 'truth' and related epistemic concepts, which results in a conflation of first- and second-order concerns.

Here's an example of the fakeness of veritism in action:

> On the contrary, truth (along with evidence, facts, and other words science studies scholars tend to relegate to scare quotes) is a far more plausible choice for one of a potential plurality of regulative ideals for an enterprise that, after all, does have an obviously cognitive function. (Baker and Oreskes 2017: 69)

The sentence prima facie commits the category mistake of presuming that 'truth' is one more – albeit preferred – possible regulative ideal of science alongside, say, instrumental effectiveness, cultural appropriateness and so on. However, 'truth' in the logical positivist sense is a feature of *all* regulative ideals of science, each of which should be understood as specifying a language game that is governed by its own validation procedures – the rules of the game, if you will – in terms of which one theory is determined (or 'verified') to be, say, more effective than another or more appropriate than another.

As an epistemic policy, veritism says that whatever else enquiry might seek to achieve in terms of the enquirers' own aims, it must first serve 'The Truth'. The result has been some strange epistemological doctrines, including 'reliabilism', which argues that there are processes that generate truths on a regular basis even if their possessors lack epistemic access to them (Goldman 1999). On the surface, such a doctrine is designed to dissociate truth from any subjective states, which would otherwise make it difficult to generalize truth claims from an individual's own version of 'justified belief', let alone personal experience. However, to the post-truther, reliabilism looks simply like a pretext for individuals to outsource their own judgements to experts in, say, cognitive, brain and/or behavioural science – those unelected masters of what remains unknown about us to ourselves. In any case, to the disinterested observer, veritism seems so keen to distance truth from utility that it resorts to a definition of truth that taken at face value is profoundly useless.

Richard Rorty became such a hate figure among analytic philosophers in the last two decades of the twentieth century because he called out the veritists on their fakeness. He observed that philosophers can tell you what truth is, but only so long as you accept a lot of contentious assumptions – and hope those capable of contending those assumptions are not in the room when you are speaking! In effect, Rorty refused to adopt any version of the 'double-truth' doctrine for philosophy comparable to the various 'double-truth' doctrines promulgated in the Middle Ages to save religious faith from critical scholarship, whereby amongst themselves philosophers adopted a semi-detached attitude towards various conflicting conceptions of truth while at the same time presenting a united front to non-philosophers, lest these masses start to believe some disreputable things.

As Rorty (1979) had explained, his own post-truth vision had been shaped by his encounter with Wilfrid Sellars's (1963) distinction between the 'manifest' and 'scientific' images of the world. Sellars's point was that the two images were 'incommensurable' in the sense that Kuhn would popularize. In other words, they cross-classify the same world for different purposes, in which case any straightforward 'reduction' or even evaluation of one image by the other is arguably an exercise in special pleading for a preferred world view. Thus, to say that a common-sense observation is contradicted by a scientific finding is to tacitly assume that the observation should be held accountable to the finding – or, more bluntly, that the ordinary person should be playing the scientist's language game. Thus, positivism – both in its original Comtean sociological and later Carnapian logical forms – always carried a strong sense of trying to reform the world. My own social epistemology is also informed by this sensibility.

Sellars's distinction influenced a range of philosophers who otherwise stand on opposite sides of key issues in epistemology and philosophy of science, including Bas van Fraassen (scientific antirealist), Paul Churchland (scientific realist) and, most relevant for our purposes, the self-styled 'anarchist' philosopher of science Paul Feyerabend (1981). Feyerabend was an avowed enemy of what he dubbed 'methodolatrists', namely, those philosophers (and scientists, of course) who place great store by particular rituals – demonstrations of methodological probity – which are said to increase the likelihood that the resulting facts enjoy the desired state of 'correspondence to reality'. As in the days of the Pharisees and the Puritans, a rigorous demeanour is made a proxy for access to truth. But Feyerabend (1975) revealed both the rhetorical strength and weakness of this 'truther' strategy by mobilizing the case of Galileo, a sloppy and perhaps fraudulent wielder of the scientific method by modern standards. Here was someone who did not understand the optics behind the 'telescope', a pimped-up periscope to the naked eye, including the eyes of his papal inquisitors. Yet, we would say that Galileo's methodologically rigorous inquisitors –perhaps by virtue of their own rigor –rendered themselves blind to the 'full truth' of his claims.

To be sure, Feyerabend leaves the moral of his story tantalizingly open. Nevertheless, the post-truther is clear that we know that Galileo was right because the rules of the science game had changed within a few decades of his death to allow his original knowledge claims to be re-established on new and improved grounds, courtesy of Newton and his followers. Galileo's interlocutors had overlooked that while failing to meet *their* standard of evidence, he was predicting something about the future of science itself that would make them obsolete and enable his knowledge claims to become facts. Galileo's sloppiness and duplicity were thus a risky epistemic investment that paid off in the long term but of course not in the short term. He was trying to play by the rules of a game other than the one to which he was being held to account. Galileo's trial displayed the difficulties of trying to change the rules of a game from within the game while the players think that the rules are fine. This last gloss helps motivate Kuhn's claim that scientists will not shift to a new paradigm until the old one has accumulated enough unsolved problems.

As we have seen, the post-truther plays two games at once: of course, she or he plays the knowledge game in which she or he is situated, in which she or he may have little prima facie 'room for manoeuvre' (*Spielraum*). But she or he also plays – at least in his or her own mind – a second and more desirable game, into which she or he would like to convert the current game. This explains that the value that the sophists placed on *kairos*, the opportunity to argue a specific case. It amounts to a search for the dialectical tipping point, a moment that the gestalt might just switch from 'duck' to 'rabbit'. In that respect, the post-truther is an epistemic 'double agent' and hence open to charges of hypocrisy in a way that the truther is not. I have associated this sense of double agency with 'bullshit', as incensed veritists have applied the term to postmodernists for nearly four decades now (Fuller 2009: chap. 4). However, a relatively neutral settling of the scores between truthers and post-truthers would conclude that post-truthers aim to weaken the fact/fiction distinction – and hence undermine the moral high ground of truthers – by making it easier to switch between knowledge games, while the truthers aim to strengthen the distinction by making it harder to switch between knowledge games. In short, the difference turns on the resolution of a struggle over what I earlier called 'modal power'.

Consensus: Manufactured Consent as the Regulative Ideal of Science?

When veritists say that truth is a 'regulative ideal' of all enquiry, they are simply referring to a social arrangement whereby the self-organizing scientific community is the final arbiter on all knowledge claims accepted by society at large. To be sure, the scientific community can get things wrong – but things become wrong only when the scientific community says so, and they become fixed only when the scientific community says so. In effect, veritists advocate what I have called 'cognitive authoritarianism' (Fuller 1988: chap. 12). From a post-truth standpoint, veritism amounts to a lightly veiled moral crusade, as exemplified by such pseudo-epistemic concepts as 'trust' and 'reliability', in which 'scientific' attaches to both a body of knowledge and the people who produce that knowledge. I say 'pseudo' because there is no agreed specifically *epistemic* measure of these qualities. Judgements about people are invariably used as proxies for judgements about the world.

For example, *trust* is a quality whose presence is felt mainly as a double absence, namely, a studied refusal to examine knowledge claims for oneself, the result of which is then judged to have had non-negative consequences – presumably because some 'trusted party' (aka scientists) did the requisite validation work. I have called trust a 'phlogistemic' concept for this reason, as it resembles the pseudo-element phlogiston (Fuller 1996). Indeed, my general opposition to this sensibility has led me to argue that universities should be in the business of 'epistemic trust-busting'. Here is my original assertion:

> *In short, universities function as knowledge trust-busters whose own corporate capacities of 'creative destruction' prevent new knowledge from turning into intellectual property.* (Fuller 2002: 47; italics in original)

By 'corporate capacities', I mean the various means at the university's disposal to ensure that the people in a position to take forward new knowledge are not simply part of the class of those who created it in the first place. Of course I had in mind ordinary teaching that aims to express even the most sophisticated concepts in terms ordinary students can understand and use, thereby deconstructing the rather historically specific – or 'path-dependent' – ways that innovations tend to become socially entrenched, which in turn create relationships of trust between 'experts' and 'laypeople'. But also I meant to include 'affirmative action' policies that are specifically designed to incorporate a broader range of people than might otherwise attend the university. Taken together, these counteract the 'neo-feudalism' to which academic knowledge production is prone – 'rent-seeking', if you will – and to which veritists are largely oblivious.

As for the veritists' core truth criterion, *reliability*, its meaning depends on specifying the conditions – say, in the design of an experiment – under which a pattern of behaviour is expected to occur. Outside of such tightly defined conditions, which is where most 'scientific controversies' happen, it is not clear how cases should be classified and counted, and hence what 'reliable' means. Indeed, STS has not only drawn attention to this fact but it has also gone further – say, in the work of Harry Collins (1985) – to question whether even lab-based reliability is possible without some sort of collusion between researchers. In other words, the social accomplishment of 'reliable knowledge' is at least partly an expression of solidarity among members of the scientific community – a closing of the ranks, to put it less charitably. This is a less flattering characterization of what veritists claim as the epistemically luminous process of 'consensus formation' in science.

An especially good example of the foregoing is what has been dubbed 'Climategate', which was triggered by the hacking of the computer server of the United Kingdom's main climate science research group at the University of East Anglia in 2009, which was followed up with several Freedom of Information requests. While no wrongdoing was formally established, the emails did reveal the extent to which scientists from across the world effectively conspired to present the data for climate change in ways that obscured interpretive ambiguities, thereby pre-empting possible appropriations by so-called 'climate change sceptics'. The most natural way to interpret this situation is that it reveals the microprocesses by which a scientific consensus is normally and literally 'manufactured'. Nevertheless, veritists are unlikely to regard Climategate as their paradigm case of a 'scientific consensus'. But why not?

The reason lies in their refusal to acknowledge the labour and even struggle that are involved in securing collective assent over any significant knowledge claim. For veritists, informed people draw the same conclusions from the same evidence. The actual social interaction among enquirers carries little cognitive weight in its own right. Instead it simply reinforces what any rational individual is capable of inferring for him- or herself in the same situation. Other people may provide additional data points, but they do not alter the rules of right reasoning. The contrasting post-truth view of consensus formation is more explicitly 'rhetorical' (Fuller and Collier 2004). It appeals to a mix of strategic and epistemic considerations in a setting where the actual interaction between the parties sets the parameters that define the scope of any possible consensus. Even Kuhn, who valorized consensus as the glue that holds together normal science puzzle solving, clearly saw its rhetorical and even coercive character, ranging from pedagogy to peer review. Let me now finally turn explicitly to the rhetorical power of consensus formation in science.

My suspicions about the presence of any sort of consensus is long-standing. The only chapter of my doctoral dissertation that appeared in my first book was on just this topic (Fuller 1988: chap. 9). My double question to anyone who wishes to claim a 'scientific consensus' on anything is *on whose authority* and *on what basis* such a statement is made. Even that great defender of science, Popper, regarded scientific facts as no more than conventions that are agreed mainly to mark temporary settlements in an ongoing collective journey. Seen with a rhetorician's eye, a 'scientific consensus' is demanded only when scientific authorities feel that they are under threat in a way that cannot be dismissed by the usual peer-review processes. 'Science' after all advertises itself as the freest enquiry possible, which suggests a tolerance for many crosscutting and even contradictory research directions, all compatible with the current evidence and always under review in light of further evidence. And to a large extent, science does demonstrate this spontaneous embrace of pluralism, albeit with the exact options on the table subject to change. To be sure, some options are pursued more vigorously than others at any given moment. Scientometrics can be used to chart the trends, which may make the 'science watcher' seem like a stock market analyst. But this is more 'wisdom of crowds' stuff than a 'scientific consensus', which is meant to sound more authoritative and certainly less transient.

Indeed, invocations of a 'scientific consensus' become most insistent on matters that have two characteristics, which are perhaps necessarily intertwined but in any case take science outside of its juridical comfort zone of peer review: (1) they are inherently interdisciplinary and (2) they are policy relevant. Think climate change, evolution, anything to do with health. A 'scientific consensus' is invoked on just these matters because they escape the 'normal science' terms in which peer review operates. To a defender of the orthodoxy, the dissenters appear to be 'changing the rules of science' simply in order to make their case seem more plausible. However, from the standpoint of the dissenter, the orthodoxy is artificially restricting enquiry in cases where reality does not fit its disciplinary template, and so perhaps a change in the rules of science is not so out of order.

Here it is worth observing that defenders of the 'scientific consensus' tend to operate on the assumption that to give the dissenters any credence would be tantamount to unleashing mass irrationality in society. Fortified by the fledgling (if not pseudo-) science of 'memetics', they believe that an anti-scientific latency lurks in the social unconscious. It is a susceptibility typically fuelled by religious sentiments, which the dissenters threaten to awaken, thereby reversing all that modernity has achieved.

Of course, there are hints of religious intent in the ranks of dissenters. One notorious example is the Discovery Institute's 'Wedge strategy', based on Phillip E. Johnson (1997), which projected the erosion of 'methodological naturalism' as the 'thin edge of the wedge' to return the United States to its Christian origins. Its very existence helped kill the defence's case for state school teaching of intelligent design in the *Kitzmiller v. Dover Area School District* case of 2005. Nevertheless, the paranoia of the orthodoxy underestimates the ability of modernity – including modern science – to absorb and incorporate the dissenters, and come out stronger for it. The very fact that intelligent design theory has translated creationism into the currency of science by leaving out the Bible entirely from its argumentation strategy should be seen as evidence for this point. And now Darwinists need to try harder to defeat it, which we see in their increasingly sophisticated refutations, which often end up with Darwinists effectively conceding points and simply admitting that they have their own way of making their opponents' points, without having to invoke an 'intelligent designer'.

In short, the very idea of a 'scientific consensus' is epistemologically over-sold. It is clearly meant to carry more normative force than whatever happens to be the cutting edge of scientific fashion this week. Yet, what is the life expectancy of the theories around which scientists congregate at any given time? For example, if the latest theory says that the planet is due for climate melt-down within 50 years, what happens if the climate theories themselves tend to go into meltdown after about 15 years? To be sure, 'meltdown' is perhaps too strong a word. The data are likely to remain intact and even be enriched, but their overall significance may be subject to radical change. Moreover, this fact may go largely unnoticed by the general public, as long as the scientists who agreed to the last consensus are also the ones who agree to the next consensus. In that case, they can keep straight their collective story of how and why the change occurred – an orderly transition in the manner of a dynastic succession in politics.

As we have seen, what holds this story together – and is the main symptom of epistemic overselling of scientific consensus – is a completely gratuitous appeal to the 'truth' or 'truth-seeking' (aka 'veritism') as somehow underwriting this consensus. Yet even the staunchest advocates of consensus thinking believe that consensus needs to be *built*. This turn of phrase comports well with the normal social constructivist sense of what consensus is. And there is nothing wrong with trying to align public opinion with certain facts and values, even on the grand scale suggested by the idea of a 'scientific consensus'. This is the stuff of politics as usual. However, whatever consensus is thereby forged – by whatever means and across whatever range of opinion – has no 'natural' legitimacy. Moreover, it neither corresponds to some pre-existent ideal of truth nor is composed of some invariant 'truth stuff' (cf. Fuller 1988: chap. 6). It is a social construction, full stop. If the consensus is maintained over time and space, it will not be due to its having been blessed or guided by 'Truth'; rather, it will be the result of the usual social processes and associated forms of resource mobilization – that is, a variety of external factors which at crucial moments impinge on the play of any game.

The idea that consensus enjoys some epistemologically more luminous status in science than in other parts of society (where it might be simply dismissed as 'groupthink') is an artefact of the routine rewriting of history that scientists do to rally their troops. As Kuhn (1970) observed, scientists exaggerate the degree of doctrinal agreement to give forward momentum to an activity that is ultimately held together simply by common patterns of disciplinary acculturation and day-to-day work practices. Kuhn's work helped generate the myth of consensus, probably with his tacit blessing. Indeed, in my Cambridge days studying history and philosophy of science with Mary Hesse (circa 1980), the idea that an ultimate consensus on the right representation of reality might serve as a transcendental condition for the possibility of scientific enquiry was highly touted, courtesy of the then fashionable philosopher Jürgen Habermas (1971), who flattered his anglophone fans by citing the early US pragmatist Charles Sanders Peirce as his source for the idea. Historical revisionism, even when done by philosophers, always begins at home.

Chapter 3

SOCIOLOGY AND SCIENCE AND TECHNOLOGY STUDIES AS POST-TRUTH SCIENCES

Sociology and the Social Construction of Identity

It is tempting to understand 'post-truth' from the standpoint of those who promoted it to become *Oxford English Dictionary*'s 2016's word of the year. As we have seen, the word pejoratively refers to those who refuse to listen to reason and evidence but instead resort to emotion and prejudice. However, to a sociologist of knowledge this formulation itself sounds too self-serving to be true. After all, the people who tend to be demonized as 'post-truth' – from Brexiteers to Trumpists – have largely managed to outflank the experts at their own game, even if they have yet to succeed in dominating the entire field of play. Empirically speaking at least, this suggests that the experts are not as rational and evidence based as they themselves thought and the post-truthers are not as emotional and prejudiced as the experts have thought them to be.

My own way of dividing the 'truthers' and the 'post-truthers' is in terms of whether one plays by the rules of the current knowledge game, or one tries to change the rules of the game to one's advantage. Unlike the truthers, who play by the current rules, the post-truthers want to change the rules. They believe that what passes for truth is relative to the knowledge game one is playing, which means that depending on the game being played, certain parties are advantaged over others. Post-truth in this sense is a recognizably social constructivist position, and many of the arguments deployed to advance 'alternative facts' and 'alternative science' nowadays betray those origins. They are talking about worlds that could have been and still could be – the stuff of modal power.

The 2017 death of the original sociological promoter of 'social construction' by name, Peter Berger, co-author of the classic *The Social Construction of Reality* (Berger and Luckmann 1966), provided an opportunity for his detractors to link him to the post-truth mentality, as he had received financial support from the US tobacco lobby in the 1970s and '80s. Indeed, he portrayed the anti-smoking campaign as primarily a moral crusade in his contribution to the landmark pro-smoking volume, *Smoking and Society* (Tollison 1985). That was because public smoking bans involve – by cause or consequence – a significant shift in how people think about public and private space, not to mention how they value the activity of smoking itself. In effect, the bans have been a game changer in terms of the norms of ordinary social interaction (cf. Collins 2004). This in turn has made the evidence for smoking's negative health effects appear more salient than whatever virtues of sociability and relaxation had been previously associated with the activity.

There is no doubt that Berger was a post-truth thinker. It enabled him to acknowledge at the outset something that latter-day defenders of the anti-smoking scientific consensus (e.g. Oreskes and Conway 2011) have been reluctant to admit, namely, that any public controversy – not least a science-based one – is simultaneously about framing the terms of the controversy and the position one adopts once the terms have been set. The extent to which smoking is problematic depends on whether the practice is understood as mainly a private lifestyle choice or a public health issue. Of course, the two interpretations are intertwined in terms of consequences.

Nevertheless, an open question in morals, politics and law remains as to who bears the burden of risk – the person who wishes to smoke or the person who does not want to be exposed to smoke. This will determine who will pay what price to enable everyone to remain part of the same society. It was a classic disagreement between the ideals of 'liberalism' and 'socialism', in which the 'socialist' framing ended up prevailing: the mass of people who reject being exposed to smoke won. Against this outcome, the losers in the second-order battle to frame the controversy – 'liberals' like Berger and his funders – are portrayed as 'motivated' to distort the truth in a way that their opponents are not. But of course, both sides are 'motivated' because they are operating from fundamentally different visions of how the world should be organized. In this regard, Berger's approach is very much of a piece with the 'cultural theory of risk' developed by the anthropologist Mary Douglas and the political scientist Aaron Wildavsky (e.g. Thompson et al. 1990).

One might say more generally that the discipline of sociology was born under the sign of 'post-truth'. After all, according to Émile Durkheim, one of the promised virtues of sociology was that it could provide 'moral education' in the rapidly changing world of the French Third Republic, in which the rules of the game switched from being based on religious and family ties to a common secular national identity, as signified by the post-1870 turnover of the educational system from religious to republican authorities. Indeed, Durkheim first practiced 'sociology' in an education faculty. It is worth observing that the other great disciplinary founder of the field, Max Weber, also lived in the generation following the consolidation of his nation in its modern form, the German Second Reich, which was itself the product of the same Franco-Prussian War that had led to the decisive shift from a Catholic monarchy to a secular republic in France.

Both Durkheim and Weber – as well as the other founders of sociology – devoted much of their energies to studying the psychosocial consequences of the rules of the game changing so quickly at the end of the nineteenth century, after having built a head of steam over the previous hundred years (Mazlish 1989). Durkheim memorably identified this phenomenon as *anomie*, namely, the rootlessness – sometimes eventuating in suicide – that results when people are caught between the two very different normative regimes: they have clearly left the old regime without quite having accommodated to the emerging new one.

Sociology in the post-truth key is mainly about the management of anomie, which is more neutrally understood as social shape-shifting. There are various styles for doing this. It can take the relatively personal character of, say, Georg Simmel on the stranger (aka 'Jew') or W. E. B. Du Bois on double consciousness (aka 'Negro'). Or, it may be seen in more abstract and general terms as ingrained in the political or professional life. Thus, Pareto stressed the need for any successful politician to decide whether to be a maintainer ('lion') or a changer ('fox') of the status quo, while Robert Merton (1976) focused on the ambivalence that professionals face in managing 'expert' and 'lay' expectations of their behaviour, where the two groups are understood to operate under partly conflicting normative regimes. Erving Goffman's dramaturgical approach to social life and ethnomethodology are probably the most self-consciously post-truth sociologies yet conceived.

However, society today is not what it was in Durkheim's – or even Goffman's – day. For the past quarter-century or more, we have witnessed the decline of the welfare state's universalist conception of society, which stabilized – at least at the official 'macro' level – the normative expectations of 'developed' societies, which defined the rules by which they were to be achieved and the milestones and goalposts of their achievement. It meant that political disagreements were conducted on relatively narrow terms by an increasingly professionalized political class. Moreover, even at the metalevel, the Cold War's defining incommensurable value systems of 'Capitalism' and 'Communism' were transacted in terms of mutually recognized league tables, ranging from size of arsenal to size of gross domestic product (GDP). Indeed, 'game theory' as we know it today in decision science was developed in this context. Truth and post-truth sensibilities seemed to merge perfectly, as the avoidance of global nuclear war was seen by all sides as the only game worth playing. At the level of science, it was the sort of convergence that would have pleased Kuhn.

The current post-truth situation could not be more different, but sociology is no less implicated. It is marked by the increasing devolution of social identity from the nation state to largely self-organizing and self-recognizing groups, each increasingly permitted its own terms of engagement with the rest of society, as manifested in proprietary control over language, space and resources. This is what is nowadays meant by 'identity politics', which informed the influential spin that Michael Burawoy (2005) gave to the mission of what he called 'public sociology' in his 2004 presidential address to the American Sociological Association. To be sure, this development has been presented negatively as fostering a 'culture of complaint', whereby one is not permitted even to refer to certain self-identified groups except according to the rules of their own language game, which group members play largely to extract concessions from the rest of society (Hughes 1993). Indeed, one need not be a cynic to suspect identity politics of encouraging a 'rent-seeking' attitude towards the social world. The more positive gloss is that identity politics constitutes a widening of social horizons, a pluralizing of the language games played, each with its own distinct objectives and associated skill sets.

Wearing its best face, identity politics presents itself as a post-truth paradise without the classically destabilizing effects of anomie, which in the end reflected the nation state's unsustainable sense of social conformity. Instead people need to be affirmed simply for who they are in their chosen field of play. Yet, we are beginning to see a sting in the tail: the 'trans' phenomenon – as in 'transgender', 'transracial' and even 'transhuman'. It takes the post-truth sensibility to its logical conclusion by assuming that in principle anyone can become adept at playing any identity game, even the most ontologically fundamental ones. In this context, transhumanists speak of 'morphological freedom', as the cornerstone of a 'Transhuman Bill of Rights' (Fuller 2016b). Perhaps it is no surprise that the transhumanist who has thought the hardest about the extent to which online identities (our 'avatars') are continuous with offline identities (our normal physical selves), Martine Rothblatt, is also the person who in transitioning from a male to a female identity became the wealthiest female CEO in the United States (Fuller and Lipinska 2016).

But if you think that a liberal attitude to gender transitioning would extend to race transitioning, then think again in light of the unprecedented controversy in the US philosophical community surrounding the publication of Rebecca Tuvel (2017), which argued that moving from, say, white to black should be judged in the same spirit as one would judge moving from male to female: if one is allowed, so too should the other. Two sorts of objections might be raised to this argument, one 'truth' and the other 'post-truth' oriented. The 'truth' one, which sparked the controversy around the article, is that in some sense gender transitioning is more 'legitimate' than race transitioning, where the meaning of 'legitimate' remains strategically vague. (I believe that this objection ultimately speaks to no more than the prejudices of our times.) The 'post-truth' objection is that identity mobility in general – be it gender, race or anything else – might instate new inequalities via an asymmetric flow of transition traffic, if, say, more men benefit from becoming women than women becoming men. And of course, above it all is the Olympian objection that to entertain such an indefinite degree of identity mobility in the first place is to take the idea of normative regimes as games much too literally. Unfortunately, that is what it means to be on the cutting edge of our post-truth world: games are serious.

Here honourable mention needs to be made of one post-truth attempt to spin identity strategically, which nevertheless in true post-truth fashion has only served to open up new fields of play. This is the concept of *intersectionality*, which was introduced by the US lawyer Kimberlé Crenshaw (1991) and became a cornerstone of 'critical race theory'. The anchoring event was the 1991 Senate hearings to confirm Clarence Thomas to the US Supreme Court. Thomas was a black conservative judge who had served in the Reagan administration, during which he allegedly sexually harassed a black female employee lawyer, Anita Hill. Hill gave extensive testimony at the Senate hearings of her experience with Thomas, to which Thomas's mainly white defenders responded that she was a tool of the white liberal establishment who wanted him not to be appointed because he did not conform to its political understanding of a black person. Thomas was eventually confirmed and continues to serve on the Supreme Court. However, in the process black identity was arguably turned into a token in a larger, largely white person's game that could be used by players on opposing sides. The 'black card' was played on Thomas's behalf against a fellow black who was portrayed as only 'pseudo-black'.

Intersectionality aims to prevent this sort of thing from happening again by reminding blacks and women that there remain systemic forms of oppression common to them as a race and a gender that transcend whatever other political and personal disagreements they might have amongst themselves. Thus, they should be careful with which other identity-based groups they align. But given the lack of agreement on whether race, class, gender or some other categorical slice of the human condition should enjoy the higher-order status that Crenshaw attributed to race and gender (in that order), intersectionality itself has become a moveable feast, a source of intense – and often acrimonious – discussions at professional sociological meetings that are then followed down the usual scholarly rabbit holes. Here the 'trans' phenomenon complicates matters tremendously, since it broadens the concerns of intersectionality from the use that might be made of already existing identities for actual strategic purpose to the acquisition of new identities for potential strategic purpose.

One context in which the problems surrounding intersectionality hit home is a very high-minded politically correct form of veritism that is currently popular among feminists and multiculturalists: *epistemic justice* – or more precisely, epistemic *in*justice (Fricker 2007). The nuance matters. One might think that 'epistemic justice' is about doing justice to *knowledge*, in response to which various theories of justice might be proposed that weigh the competing demands of equality, fairness, desert, cost, benefit and so forth in the production, distribution and consumption of knowledge. This captures the post-truth spirit of my own 'social epistemology' because it presumes that the norms governing knowledge and knowers need to be forged simultaneously: who knows and what is known are always mutually implicated. Identities are not fixed in stone – or in the genetic or historical record, for that matter. Epistemic justice should be about deciding these matters at both the individual and the collective level.

In contrast, 'epistemic injustice' in its fashionable form presumes that an unproblematic conception of 'social justice' sets the rules of the epistemic game so that the normative objective is to identify and correct violations of play, hence 'epistemic *in*justice'. These injustices typically relate to members from clearly marked socially discriminated groups who are prevented from contributing their unique perspectives, experiences and data to an already socially agreed sense of enquiry. And without denying the historic neglect of women and minority voices, which has damaged both them and any greater sense of human enquiry, one wonders whether this strict sense of 'epistemic injustice' can survive the ongoing 'trans' revolution. Nearly twenty years ago I considered something similar in terms of whether 'affirmative action' policies in universities should be understood as primarily promoting, say, women as individuals, regardless of their epistemic dispositions, or a distinctive form of 'women's knowledge', regardless of the individuals, including men, who might bear it (Fuller 2000a: chap. 4). Of course, one might wish to promote both senses of affirmative action, as I am inclined to do, but they require different policies that may interestingly cut against each other, once implemented.

Science and Technology Studies and Scientific Gamesmanship

Post-truth is the offspring that science and technology studies (STS) has been always trying to disown, not least in a recent editorial of the field's main journal, *Social Studies of Science* (Sismondo 2017). Yet STS can be fairly credited with having both routinized in its own research practice and set loose on the general public – if not outright invented – at least four common post-truth tropes:

1. Science is what results once a scientific paper is published, not what made it possible for the paper to be published, since the actual conduct of research is always open to multiple countervailing interpretations.
2. What passes for the 'truth' in science is an institutionalized contingency, which if scientists are doing their job will be eventually overturned and replaced, not least because that may be the only way they can get ahead in their fields.
3. Consensus is not a natural state in science but one that requires manufacture and maintenance, the work of which is easily underestimated because most of it occurs offstage in the peer-review process.
4. Key normative categories of science such as 'competence' and 'expertise' are moveable feasts, the terms of which are determined by the power dynamics that obtain between specific alignments of interested parties.

Yet, STS talks the talk without ever quite walking the walk. What is puzzling, especially from a strictly epistemological standpoint, is that STS recoils from these tropes whenever such politically undesirable elements as climate change deniers or creationists appropriate them effectively for their own purposes. Normally, that would be considered 'independent corroboration' of the tropes' validity, as these undesirables demonstrate that one need not be a politically correct STS practitioner to wield the tropes effectively. In this respect, STS practitioners seem to have forgotten the difference between the contexts of discovery and justification in the philosophy of science. The undesirables are actually helping STS by showing the robustness of its core insights, since these are people who otherwise overlap little with the normative orientation of most STS practitioners, yet that have managed to turn those insights to good effect, at least as far as the undesirables themselves are concerned (Fuller 2016a).

Of course, STS practitioners are free to contest any individual or group that they find politically undesirable. But that is a political not a methodological battle. STS should not be quick to fault undesirables for 'misusing' its insights, let alone apologize for, self-censor or otherwise restrict the field's practitioners who apply those insights. On the contrary, it should defer to the wisdom of Oscar Wilde and simply grant that imitation is the sincerest form of flattery. STS has enabled the undesirables to raise their game, and if STS practitioners are too timid to function as partisans in their own right, they could try to help the desirables raise their game in response.

Take the ongoing debates surrounding the teaching of evolution in the United States. The fact that intelligent design theorists are not as easily defeated on scientific grounds as young earth creationists means that when their Darwinist opponents leverage their epistemic authority as if the intelligent design theorists were 'mere' creationists, the politics of the situation becomes naked. Thus, unlike previous creationist high court cases, the judge in *Kitzmiller v. Dover Area School District* – the 2005 case in which this author served as an expert witness for the defence – dispensed with the niceties of the philosophy of science and resorted to the brute sociological fact that most evolutionists do not consider intelligent design theory science. That was enough for the Darwinists to win the battle, but will it win them the war?

Those who have followed the 'evolution' of creationism into intelligent design over the past 40 years might conclude that Darwinists act in bad faith because they do not take seriously that intelligent design theorists try to play by the Darwinists' rules by highlighting scientifically admitted lacunae in the neo-Darwinian framework. Indeed, a dozen years after *Kitzmiller*, there is little evidence that Americans are any friendlier to Darwin than they were before the trial. And no one thinks that Trump's residency in the White House will change matters soon.

STS officially recoiled from the post-truth world view in 2004, when Bruno Latour famously waved the white flag in the 'Science Wars', which had been already raging for nearly fifteen years – an intellectual by-product of the post-Cold War reassessment of public funding for science, in which STS-style demystification played a noticeable role (Ross 1996; cf. Fuller 2006b). The terms of surrender in Latour (2004) are telling. Latour basically issued a mea culpa on behalf of STS for having appeared more deconstructive of the scientific establishment than the field had intended. Properly understood, so says the chastened Latour, STS is simply the empirical shadow of the fields it studies. It basically amplifies consensus where it exists, showing how it has been maintained, and amplifies dissent where it exists, similarly showing how it has been maintained. Here Latour's famous methodological exhortation to 'follow the actors' clings tightly to his mentor Michel Serres's positive estimation of the *parasite* as a role model (Serres and Latour 1995; cf. Fuller 2000b: chap. 7). If STS seems 'critical', that is only an unintended consequence of the many policy issues involving science and technology which remain genuinely unresolved. STS adds nothing to settle the normative standing of these matters. It simply elaborates them and in the process perhaps reminds people of what they might otherwise wish to forget or ignore.

This was quite a climb down for Latour. After all, in the book that launched him to intellectual superstardom, *We Have Never Been Modern*, Latour (1993) attempted to universalize what the founding Edinburgh School of STS had dubbed the *symmetry principle*, a methodological tenet associated with a broadly 'naturalistic' approach to social enquiry, whereby the same sorts of causal factors would be invoked when explaining episodes in the history of science and technology, regardless of whether we now deem them to have been 'good' or 'bad' (Bloor 1976). Latour's seemingly bold move, which has continued to make him the darling of 'post-humanists', was to extend this principle so that not only are our value preferences not privileged in social explanations but also any sense of species chauvinism, which would have humans take exclusive credit for bringing about states of the world that required the participation of non-human beings.

However, Latour had not anticipated that symmetry applied not only to the range of objects studied but also the range of agents studying them. Somewhat naively, he seemed to think that a universalization of the symmetry principle would make STS the central node in a universal network of those studying 'technoscience'. Instead, everyone with access to the writings of Latour and other STS practitioners started to apply the symmetry principle for themselves, which led to rather cross-cutting networks and unexpected effects, especially once the principle started to be wielded by creationists, climate sceptics, New Age medics and other candidates for an epistemic 'basket of deplorables', in Clinton's notorious phrase. And by turning symmetry to their advantage, the deplorables got results, at least insofar as the balance of power has gradually tilted more in their favour – again, for better or worse.

I believe that a post-truth world is the inevitable outcome of greater epistemic democracy. In other words, once the instruments of knowledge production are made generally available – and they have been shown to work – they will end up working for anyone with access to them. This in turn will remove the relatively esoteric and hierarchical basis on which knowledge has traditionally acted as a force for stability and often domination. As we have seen, the locus classicus for esotericism is the *Republic*, in which Plato promotes what in the Middle Ages started to be called a 'double truth' doctrine – one for the elites (which allows them to rule) and one for the masses (which allows them to be ruled).

Of course, the cost of making the post-truth character of knowledge so visible is that it also exposes a power dynamics that may become more intense and ultimately destructive of the social order. As we have seen, this was Plato's take on democracy's endgame. In the early modern period, this first became apparent with the Wars of Religion that almost immediately broke out in Europe once the Bible was made readily available. Bacon and others saw in the scientific method a means to contain any such future conflict by establishing a new epistemic mode of domination. While it is possible to defer democracy by trying to deflect attention from the naked power dynamics à la

Latour with fancy metaphysical diversions and occasional outbursts in high dudgeon, those are leonine tactics that only serve to repress STS's foxy roots. STS should finally embrace its responsibility for the post-truth world and call forth the field's vulpine spirit to do something unexpectedly creative with it. After all, the hidden truth of *Aude sapere* (Kant's 'Dare to know') is *Audet adipiscitur* (Thucydides's 'Who dares, wins').

Nevertheless, STS remains bound to the Latourian surrender. This leads Sergio Sismondo (2017), the editor of the field's main journal, to extol the virtues of someone who seems completely at odds with the STS sensibility, namely, Naomi Oreskes, the Harvard science historian turned public defender of the scientific establishment. A signature trope of her work is the pronounced *asymmetry* between the *natural* emergence of a scientific consensus and the *artificial* attempts to create scientific controversy (Oreskes and Conway 2011; Baker and Oreskes 2017). It is precisely this 'no science before its time' sensibility that STS has been supposedly spending the last half-century trying to oppose. Even if Oreskes's political preferences tick all the right boxes from the standpoint of most STS practitioners, she has methodologically cheated by presuming that the 'truth' of some matter of public concern most likely lies with what most scientific experts think at a given time. Indeed, Sismondo's passive aggressive agonizing in his editorial comes from having to reconcile his intuitive agreement with Oreskes and the contrary thrust of most STS research.

The most important issue relevant to STS addressed by post-truth is distrust in *expertise*, to which STS has undoubtedly contributed by circumscribing the prerogatives of expertise. Sismondo fails to see that even politically mild-mannered STS practitioners like Harry Collins and Sheila Jasanoff do this in their work. Collins is mainly interested in expertise as a form of knowledge that other experts recognize as that form of knowledge, while Jasanoff is clear that the price that experts pay for providing trusted input to policy is that they agree not engage in imperial overreach (e.g. Collins and Evans 2007; Jasanoff 1990). Neither position approximates the much more authoritative role that Oreskes would like to see scientific expertise play in policymaking. From an STS standpoint, those who share Oreskes's normative orientation to expertise should be thinking about how to improve science's public relations, including proposals for how scientists might be socially and materially bound to the outcomes of policy decisions taken based on their advice.

When I say that STS has forced both more and less established scientists to 'raise their game', I am alluding to what may turn out to be STS's most lasting contribution to the general intellectual landscape, namely, to think about science as literally a *game* – perhaps the biggest game in town. Consider football, where matches typically take place between teams with divergent resources and track records. Of course, the team with the better resources and track record is favoured to win, but sometimes it loses and that lone event can desta-bilize the team's confidence, resulting in further losses and even defections. Each match is considered a free space where for 90 minutes the two teams are presumed to be equal, notwithstanding their vastly different histories. Bacon's ideal of the 'crucial experiment', so eagerly adopted by Popper, relates to this sensibility as definitive of the scientific attitude. And STS's 'social constructiv-ism' simply generalizes this attitude from the lab to the world. Were STS to embrace its own sensibility much more wholeheartedly, it would finally walk the walk.

A great virtue of the game idea is its focus on the reversibility of fortunes, as each match matters, not only to the objective standing of the rival teams but also to their subjective sense of momentum. Yet, from their remarks about intelligent design theory, Baker and Oreskes (2017) appear to believe that the science game ends sooner than it really does: after one or even a series of losses, a team should simply pack it in and declare defeat. Here it is worth recalling that the existence of atoms and the relational character of space-time – two theses associated with Albert Einstein's revolution in physics – were controversial if not deemed defunct for most of the nineteenth century, not-withstanding the problems that were acknowledged to exist in fully redeeming the promises of the Newtonian paradigm. And scientists who continued to uphold such unfashionable views, such as Ernst Mach in the case of the rela-tional character of space-time, were seen as cranks who focussed too much on the lost futures of past science. Yet after the revolutions in relativity and quantum mechanics, Mach's reputation flipped and he became known for his prescience. Thus, the Vienna Circle that spawned the logical positivists was originally named in Mach's honour (Feuer 1974).

Similarly intelligent design may well be one of those 'controversial if not defunct' views that will be integral to the next revolution in biology, since even biologists whom Baker and Oreskes probably respect admit that there are serious explanatory gaps in the neo-Darwinian synthesis. That intelligent design advocates have improved the scientific character of their arguments from their creationist origins – which I am happy to admit – is not something for the movement's opponents to begrudge. Rather it shows that they learn from their mistakes, as any good team does when faced with a string of losses. Thus, one should expect an improvement in performance. Admittedly these matters become complicated in the US context, since the Constitution's sepa-ration of church and state has been interpreted in recent times to imply the prohibition of any teaching material that is motivated by specifically religious interests, as if the Founding Fathers were keen on institutionalizing the genetic fallacy! Nevertheless, this blinkered interpretation allows Baker and Oreskes to continue arguing with earlier versions of 'intelligent design creationism', very much like generals whose expertise lies in having fought the previous war. But luckily, an increasingly informed public is not so easily fooled by such epistemically rearguard actions.

Some STS scholars grant that the game metaphor is appropriate to how science is conducted today but reject my endorsement of the metaphor, especially as something that STS might apply in its own practice. Amanda Phillips (2017) interestingly discusses the matter in terms of the introduction of the mortar kick into US football, which stays within the rules of the game but threatens player safety. This leads her to conclude that the mortar kick debases/jeopardizes the spirit of the game. I may well agree with her on the specific point, which she then wishes to extend to a normative stance appropriate to STS. However, I may agree here too, but I would first like to see whether she would have disallowed past innovations that changed the play of the game – and, if so, which ones. In other words, I would need a clearer sense of what she takes to be the 'spirit of the game', which involves inter alia judgements about tolerable risks over a period of time.

Judicial decisions normally address these matters. Sometimes judges issue 'landmark decisions' which may invalidate previous judges' rulings, but in any case they set a precedent on the basis of which future decisions should be made. Bringing it back to the case at hand, Phillips might say that football has been violating its spirit for a long time, and that not only should the mortar kick be prohibited but so too some other earlier innovations. (In US Constitutional law, this would be like the history of judicial interpretation of citizen rights following the passage of the Fourteenth Amendment, at least starting with the 1954 Supreme Court case *Brown v. Board Education*, which overruled previous court rulings that interpreted 'equal' rights as permitting racial segregation.) Of course, Phillips might instead give a more limited ruling that simply claims that the mortar kick is a step too far in the evolution of the game, which so far has stayed within its spirit. Or, she might simply judge the mortar kick to be within the spirit of the game, full stop. The arguments used to justify any of these decisions would be an exercise in elucidating what the 'spirit of the game' means.

All these considerations force us to think more deeply about what it means to characterize science as a 'game'. It would seem to mean, at the very least, that science is prima facie an autonomous activity in the sense of having clear boundaries. Just as one knows when one is playing or not playing football, one knows when one is or is not doing science. Of course, the impact that has on the rest of society is an open question. For example, once dedicated schools and degree programmes were developed to train people in 'science' (and here I mean the term in its academically broadest sense, *Wissenschaft*), especially once they acquired the backing and funding of nation states, science became the source of ultimate epistemic authority in virtually all policy arenas. This was something that really only began to happen in earnest in the second half of the nineteenth century.

Similarly one could imagine a future history of football, perhaps inspired by the modern Olympics, in which larger political units acquire an interest in developing the game as a way of resolving their own standing problems that might otherwise be handled with violence, sometimes on a mass scale. In effect, the Olympics would be a regularly scheduled, sublimated version of a world war. In that possible world, football – as one of the represented sports – would come to perform the functions for which armed conflict is now used. Here sports might take inspiration from the various science 'races' in which the Cold War was conducted – notably the race to the moon – was a highly successful version of this strategy in real life, as it did manage to avert a global nuclear war. Its intellectual residue is something that we still call 'game theory'.

Baker and Oreskes (2017) correctly pick up on the analogy drawn by David Bloor (1976) between social constructivism's scepticism with regard to transcendent conceptions of truth and value and the scepticism that the Austrian school of economics (and most economists generally) show to the idea of a 'just price', understood as some normative ideal that real prices should be aiming towards. Indeed, there is more than an analogy here. Alfred Schutz, teacher of Peter Berger and Thomas Luckmann of Berger and Luckmann (1966) fame, joined Friedrich Hayek as a member of Ludwig Mises's circle in interwar Vienna, having been trained by Mises in the law faculty (Prendergast 1986). Market transactions clearly provided the original template for the idea of 'social construction', a point that is already clear in Adam Smith and became increasingly clear over the course of Berger's career (Fuller 2017), and that continues to be pursued with considerable vigour by the economic historian Deirdre McCloskey, most recently in her phenomenological history of the commercial life (McCloskey 2006, 2010).

However, in criticizing Bloor's analogy, Baker and Oreskes miss a trick: when the Austrians and other economists talk about the normative standing of real prices, their understanding of the market is somewhat idealized; hence, one needs a phrase like 'free market' to capture it. This point is worth bearing in mind because it amounts to a competing normative agenda to the one that Baker and Oreskes are promoting. With the slow ascendancy of neo-liberalism over the second half of the twentieth century, that normative agenda became clear – namely, to *make* markets free so that real prices can prevail.

Here one needs to imagine that in such a 'free market' there is a direct correspondence between increasing the number of suppliers in the market and the greater degree of freedom afforded to buyers, as that not only drives the price down but also forces buyers to refine their choice. This is the educative function performed by markets, an integral social innovation in terms of the Enlightenment mission advanced by Smith, Condorcet and others in the eighteenth century (Rothschild 2002). Markets were thus promoted as efficient mechanisms that encourage learning, with the 'hand' of the 'invisible hand' best understood as that of an instructor. In this context, 'real prices' are simply the actual empirical outcomes of markets under 'free' conditions. Contra Baker and Oreskes, they do not correspond to some a priori transcendental realm of 'just prices'.

However, markets are not 'free' in the requisite sense as long as the state strategically blocks certain spontaneous transactions, say, by placing tariffs on suppliers other than the officially licensed ones or by allowing a subset of market agents to organize in ways that enable them to charge tariffs to outsiders who want access. In other words, the free market is not simply about lower taxes and fewer regulations. It is also about removing subsidies and preventing cartels. It is worth recalling that Smith wrote *The Wealth of Nations* as an attack on 'mercantilism', an economic system not unlike the 'socialist' ones that neoliberalism has tried to overturn by appealing to the 'free market'. In fact, one of the early neo-liberals (aka 'ordo-liberals'), Alexander Rüstow, coined the phrase 'liberal interventionism' in the 1930s for the strong role that he saw for the state in freeing the marketplace, say, by breaking up state-protected monopolies (Jackson 2009).

Capitalists defend private ownership only insofar as it fosters the commodification of capital, which in turn, allows trade to occur. Capitalists need not be committed to a land-oriented approach to private property, à la feudalism, which through, say, inheritance laws restricts the flow of capital in order to stabilize the social order. To be sure, capitalism requires that traders know who owns what at any given time, which in turn supports clear ownership signals. However, capitalism flourishes only if the traders are inclined to part with what they already own to acquire something else. After all, wealth cannot grow if capital does not circulate. The state thus serves capitalism by removing the barriers that lead people to accept too easily their current status as an adaptive response to situations that they regard as unchangeable. This helps explain why liberalism, the movement most closely aligned with the emerging capitalist sensibility, was originally called 'radical' – from the Latin for 'root' (Halevy 1928). It promised to organize society according to humanity's fundamental nature, the full expression of which was impeded by existing (typically property-based) regimes, which failed to allow for everyone what the twentieth century would call 'equal opportunity' in life, not least the opportunity to trade their past for a ticket to a possibly better future.

I offer this more rounded picture of the normative agenda of free market thinkers because Baker and Oreskes engage in a rhetorical sleight of hand associated with the capitalists' original foes, the mercantilists. It involves presuming that the public interest is best served by state authorized producers (of whatever). Indeed, when one speaks of the early modern period in Europe as the 'Age of Absolutism', this elision of the state and the public is an important part of what is meant. True to its Latin roots, the 'state' is the anchor of stability, the stationary frame of reference through which everything else is defined. Here one immediately thinks of Newton, but metaphysically more relevant was Hobbes, whose absolutist conception of the state aimed to incarnate the Abrahamic deity in human form, the literal body of which is the body politic.

Setting aside the theology, mercantilism in practice aimed to reinvent and rationalize the feudal order for the emerging modern age, one in which 'industry' was increasingly understood as not a means to an end but an end in itself – specifically, not simply a means to extract the fruits of nature but an expression of human flourishing. Thus, political boundaries on maps started to be read as the skins of superorganisms, which by the nineteenth century came to be known as 'nation states'. In that case, the ruler's job is not simply to keep the peace over what had been largely self-managed tracts of land, but rather to 'organize' them so that they might function as a single productive unit, what we now call the 'economy'. The first modern theorization was as a 'physiocracy', which envisaged governing as tantamount to performing as a physician on the body politic.

The original mercantilist policy involved royal licenses that assigned exclusive rights to a 'domain' understood in a sense that was not restricted to tracts of land, but extended to wealth production streams in general. To be sure, over time these rights were attenuated into privileges and subsidies, which allowed for some competition but typically on an unequal basis. The downstream beneficiaries of this development have been academic disciplines conceptualized as 'domains', which by the late nineteenth century was institutionalized by extending the 'doctor' degree from its original administrative functions to the medieval liberal arts subjects and their modern descendants in the sciences (Fuller 2013). It is here that we begin to see the 'rent-seeking' practices that characterize the academic knowledge system today, which in the guise of the search for truth ends up utilizing all the available knowledge. I shall follow through on this line of thought in the next chapter.

In contrast to mercantilism, capitalism's 'liberal' sensibility was about repurposing the state's power to prevent the rise of new 'path dependencies' in the form of, say, a monopoly in trade based on an original royal license renewed in perpetuity, which would only serve to reduce the opportunities of successive generations. It was an explicitly anti-feudal policy. The final frontier to this policy sensibility is academia, which has long been acknowledged to be structured in terms of what Merton originally called the principle of 'cumulative advantage', the sources of which are manifold and, to a large extent, mutually reinforcing (Merton 1968b). To list just a few: (1) state licenses issued to knowledge producers, starting with the Charter of the Royal Society of London, which provided a perpetually protected space for a self-organizing community to do as they will within originally agreed constraints; (2) Kuhn-style paradigm-driven normal science, which yields to a successor paradigm only out of internal collapse, not external competition; (3) the anchoring effect of early academic training on subsequent career advancement, ranging from jobs to grants; (4) the evaluation of academic work in terms a peer-review system whose remit extends beyond catching errors to judging relevance to preferred research agendas; (5) the division of knowledge into 'fields' and 'domains', which supports a florid cartographic discourse of 'boundary work' and 'boundary maintenance'.

The list could go on, but the point is clear to anyone with eyes to see: academia continues to present its opposition to both neo-liberalism and neo-populism in the sort of neo-feudal terms that would have pleased a mercantilist. Lineage is still everything, whatever the source of ancestral entitlement. Merton's own attitude towards academia's multiple manifestations of 'cumulative advantage' seemed to be one of ambivalence, though as a sociologist he probably was not sufficiently critical of the pseudo-liberal spin put on cumulative advantage as the expression of the knowledge system's 'invisible hand' at work. However, as Alex Csiszar (2017) has recently shown, even Merton recognized that the introduction of scientometrics in the 1960s – in the form of the Science Citation Index – made academia susceptible to a tendency that he had already identified in bureaucracies, 'goal displacement', whereby once a qualitative goal is operationalized in terms of a quantitative indicator, there is an incentive to work towards the indicator, regardless of its actual significance for achieving the original goal. Thus, the cumulative effect of high citation counts becomes a surrogate for 'truth' or some other indicator-transcendent goal. In this real sense, what is at best the wisdom of the scientific crowd ends up being routinely mistaken for an epistemically luminous scientific consensus.

Chapter 4

THE POST-TRUTH ABOUT ACADEMIA: UNDISCOVERED PUBLIC KNOWLEDGE

Introduction: Academia's Epistemic Shortfalls and Entitlement Pretensions

'Academic freedom' may be the pursuit of truth wherever it may lead, but it is not obvious that left to their own devices academics will necessarily explore, let alone exploit, all that is knowable to the fullest extent. Put more provocatively, the university is inclined to compromise its own liberal universalism, unless it is compelled by external forces to alter its default patterns of behaviour. Perhaps the two most historically important countervailing forces to the university's tendency to compromise – the external drivers of academic universalism, as it were – came together in the twentieth century in what US President Dwight Eisenhower called, albeit with pejorative intent, 'the military-industrial complex'. The bulk of this chapter deals with this matter.

But right at the outset, my claim is both controversial and counterintuitive. After all, the military and industrial sectors of society are generally portrayed as inhibitors or distorters of pure academic enquiry. However, this depiction is misleading, though it has skewed the narratives that historians have told about the development of the sciences. It has led to a neglect of what US National Science Foundation (NSF) programme officer Donald Stokes (1997) originally called 'Pasteur's Quadrant', namely, basic research that is *use-inspired* – specifically, major intellectual breakthroughs that result from addressing large-scale or long-term practical concerns. The post-truth research agenda dwells here, one which distinctly stands against business as usual in academia.

The phrase 'Pasteur's Quadrant' recalls that Louis Pasteur's enduring achievements in what is now called microbiology involved bringing together knowledge from different academic disciplines, typically by challenging one or more of their fundamental assumptions. As witnessed in Pasteur's lifelong 'Franco-Prussian' rivalry with Robert Koch, his efforts were in aid of solving major practical problems relating to the national interest in commerce and

war. To be sure, these efforts are claimed by academia today as having been
'biomedical' in nature. Here the normative term 'disease' serves as a pivot
to convert Pasteur's achievements from 'applied' to 'basic' research, and his
original attempts to stop bacteria from destroying the silk, milk, wine and beer
industries or killing troops in the field are seen as proper 'scientific discoveries'
in the disciplines of biology and medicine.

Notwithstanding the academic co-optation of Pasteur's work, Stokes's
original insight stands. The guiding epistemological intuition is that the persistence of *deep* practical problems (i.e. ones that address the human condition as such and do not simply take care of themselves over time or can
be resolved in endlessly ad hoc ways) are mainly about the relatively poor
organization of academic knowledge, which in its default state prevents the
right connections from being made – as opposed to anything mysterious in
reality itself. This is the sense in which what I shall later call the 'military-
industrial will to knowledge' serves as an antidote to the epistemic shortfalls
of academic freedom.

In the library and information science literature, the failure by academics
to connect their own literatures properly to maximum benefit results in large
amounts of 'undiscovered public knowledge' (Swanson 1986). The phrase
originally referred to the ability of a library scientist to read across cognitively related literatures in two disciplines that normally ignored each other
in order to propose a viable hypothesis for solving a long-standing medical
problem – all without having first to secure a research grant to reinvent what
had already been implicitly known, were anyone bothered to search (Fuller
2016a: chap. 3). This is reflected in the well-known fact that in all fields an
overriding amount of attention is lavished on a relatively small portion of the
scholarly literature in any given field, something along the lines of Pareto's
'80/20 principle', whereby 80 per cent of the actual effects are due to 20
per cent of the available causes. There are many ways to think about this
proposition, which Pareto himself believed had very wide applicability. At the
very least, as we shall shortly see, it can be understood in either 'leonine' or
'vulpine' terms.

In any case, in the context of academic practice, it basically means that 80
per cent of academic effort is focused on 20 per cent of the available knowledge. This was one of the issues originally examined by Derek de Solla Price
(1963), the historian of science who popularized the phrase 'Big Science' to
describe our era in the course of founding the field of 'scientometrics', the
quantitative study of science's various social and economic dimensions, both
as a system in its own right and as part of larger socio-economic systems.
Price found that as more scientists entered a field and its research horizon
became more tightly focused, demonstrating a 'cutting edge', Pareto's ratio

could easily shift to 90/10. And given the way scientists cite each other's work, an increasing number of them end up becoming dependent on diminishingly few of their colleagues. The result, bluntly put, is that the vast majority of research remains uncited and typically unread – and hence without any palpable impact whatsoever.

Some see such research as increasing the 'noise' of the knowledge system, in the sense of interfering with the 'signals' sent by the worthwhile research. In this spirit Merton (1968) suggested that it would be wise for researchers who find themselves at the academic periphery to maximize the impact of their ideas by donating them to researchers nearer the core. Others – typically economists – see the research surfeit as 'noise' but in the more Panglossian sense of providing a gestalt-like 'ground' against which the worthy epistemic 'signals' can appear as 'figures'. The noise may be seen as false starts or precursors of those who end up receiving the lion's share of the citations, having been forced to raise their game precisely due to the relative scarcity of collective attention space. This was part of Pareto's own original understanding of the 80/20 ratio as a sociological law, which was inspired by Gottfried Wilhelm von Leibniz's principle of sufficient reason. It allowed Pareto to refer to the elite 20 per cent who have owned 80 per cent of the wealth over history as an 'aristocracy', albeit one whose merit is demonstrated less in terms of hereditary entitlement than mere comparative advantage.

This last point is not trivial because Pareto appeared to believe that the 80/20 principle was tantamount to a natural law for humans, insofar as no matter where a society begins and what route it takes, it ends up pursuing the same general trajectory or tendency as other societies. In general systems theory, this is called 'equifinality', which was associated with a strong fatalist sensibility in ancient times but also a strong progressivist sensibility in modern times (Fuller 2015: chap. 6). This made Pareto very difficult to read politically in his own day, though he is fairly regarded as a kind of 'liberal'.

My own considered view is that Pareto was not so far from the British Fabians and others who believed that 'elites' in some sense are endemic to the human condition, with the interesting political question being the basis on which elites are to be formed. Fabians, partly inspired by Plato and partly by the eugenicist Francis Galton, believed the basis should be 'merit', understood as a capacity for achievement in some socially relevant sphere of activity, not so very far from the psychometric understanding of 'intelligence' that was coming into fashion in the early twentieth century. On this basis, society as a whole may then be rationally organized. Where Pareto would clearly diverge from the Fabians was in thinking that the result could ever be a truly 'egalitarian' society, either in thought or deed. For Pareto, that was the ultimate hypocrisy of socialism. Indeed, Pareto's open disavowal of egalitarianism made him an awkward figure for sociologists to discuss after the incorporation of Marx into the sociological canon in the 1960s.

Nevertheless, even on Pareto's original terms, the 80/20 principle can be interpreted in ways that favour either the lion or the fox. A good way to think about the difference here is in terms of contrasting attitudes to the land-based economy in late eighteenth- and early nineteenth-century Britain, that crucible of classical political economy, which Karl Polanyi (1944) famously called the 'great transformation'. The reader should keep this in the back of their mind when assessing the strengths and weaknesses of academia's default knowledge production regime.

The lions in Polanyi's account are the Tory landowners whose paternalism was open and full-blown, all done in the name of their stewardship of God's earthly estate. They determined what portion of the land should be cultivated or lie fallow, as well as what constituted a fair portion of the yield to those who worked on it. In years where land yield fell under expectations, the harvest would be redistributed so that the workers did not starve. The landowners effectively meted out God's justice on earth. Polanyi looked upon them much more benignly than has been the norm in the modern period, which tends towards what Herbert Butterfield had dubbed in Polanyi's day as 'Whig history', named for the Tories' political opponents in Parliament. The Whigs portrayed the gradual erosion of the hereditary entitlements underwriting Tory stewardship – which by the early nineteenth century had resulted in a free labour market and democratic representation – as part of the onward march of human liberty.

In Polanyi's account, the foxes are the Whigs and their 'Liberal' successors, the champions of classical political economy, whose normative sensibility was tied to external standards, notably laws, epitomized in the definition of a republic as 'an empire of laws, and not of men' put forward by US Founding Father John Adams. While Adams's words are now uttered in reverence, Polanyi's point is that at the time such things were first said, they were understood as calling for a more game-like approach to governance, whereby potentially anyone can get involved if they are willing and able to play by the rules. One need not to have descended from a specific land-based lineage, as the Tory world view stipulated. At one level, this was certainly liberating. But it was equally risky for those would enter the field of play, since they often had to rely on little more than their wits, which might not be enough if or when they lost. The point was dramatized in Marx's concept of exploitation, which reflects how capitalists were able to play the 'laws' of political economy to their advantage against the workers. Not surprisingly, in this brave new foxy regime, the legal profession grew in stature and sophistication to become the dominant force in modern politics. For example, more than half of US presidents have been lawyers.

Before turning to academic knowledge production, a brief remark is in order about the contrasting normative sensibilities presumed by the Tory and Whig sides of the argument, which will help orient the reader to what follows. The Tory world view presumes a relatively sanguine view of the human condition, in which everyone can flourish in their own way if each recognizes the proper place of the other. Insofar as any sense of 'equality' is implied here, it is one of mutual respect – of the sort that kept the lord-serf relationship intact for centuries. On this view, all the problems that humans experience result from a temporary disequilibrium of this system of co-dependency, one which restores itself in due course. In contrast, the Whig world view presumes a less trusting approach to the human condition. From this standpoint, what really keeps the Tory peace is the threat of force, which, say, a lord could normally apply disproportionally to an offending serf. Implied here – and quite openly discussed in the *Federalist* papers that underwrote the US Constitution – is that people cannot be trusted to self-govern without considerable legal scaffolding. But once that scaffolding is in place, they are then free to play the system to their advantage – as Trump has been trying to do, with some limited success to date.

Now let us shift gears to think about academic knowledge production. Here the lions are those who defend academia's virtualized sense of hereditary lineage through alma mater ('nurturing mother') and *Doktorvater* ('doctor father'), whereby those who descend from the best schools and the best teachers are entitled to be taken more seriously in terms of both their potential and their output as knowledge producers. But now suppose that a successful academic appears to have escaped this highly path-dependent system of what Merton (1968) rightly dubbed 'cumulative advantage' – say, by having studied at a marginal university. Such a person is treated as an anomaly to be explained away by, say, their postdoctoral appointments and other incidental affiliations with institutions more closely aligned with the prescribed order of things. Indeed, in light of such explanations, the status quo is reinforced and celebrated Tory-like for its 'self-corrective' capacity.

Consider the relatively limited sense in which academics talk about 'research impact'. It is mainly restricted to their publishing in certain 'high-impact' journals that assure a quick and wide uptake. In the second half of the twentieth century great stress was placed on this process happening in a somewhat mysterious epistemological version of an 'invisible hand'. Thinkers as diverse as the economist Friedrich Hayek, the philosopher-chemist Michael Polanyi (Karl's brother) and the sociologist Robert Merton advanced such a view largely to contrast it with the spectre of totalitarianism. This led them to regard the very idea of top-down research planning as the royal road to distortion, falsification and indoctrination.

Yet seen from a post-truth perspective, research impact of the sort valorized by these would-be liberals is already 'spontaneously planned', and not necessarily in a good way, as the gatekeepers of academic knowledge production are oriented primarily to each other through long-established channels of communication and recognition. We thus arrive at the open secret of 'peer review', which is that journal editors, funding agencies and evaluation panels routinely deliver alloyed – some might say corrupt – judgements that cover both the strict validity of knowledge claims and their relevance to various preferred research agendas. The overall result is less a free marketplace of ideas than a queered pitch for the reception of new knowledge (Fuller 2016a: chap. 2).

At this point, the foxes enter to champion neglected research as an untapped resource, something that should be exploited for its hidden potential to advance already existing agendas and initiate new ones. The signature knowledge management practice of 'data mining' is one largely privatized way of pursuing this idea. 'Data surfacing' is another, and to my mind preferable, practice, albeit equally privatized: whereas the former sticks to client-specified needs, the latter allows the knowledge manager to reconfigure clients to enable them to deal more effectively with all the data on offer (Fuller 2016a: chap. 3).

But there is no reason why even nationally funded research funding agencies should not pursue an energetic academic knowledge exploitation strategy, perhaps in the spirit of 'affirmative action', especially if we already agree that academia's current impact-making mechanisms are unjustly biased along various path-dependent dimensions. In this context, the Internet may be already helping the foxes' agenda along, as a more educated population increasingly channels its enquiries through Internet search engines, as opposed to such usual academic gatekeeping agencies as textbooks and certified experts. These search engines serve to democratize the presentation of knowledge by virtue of their list-like formatting alone, notwithstanding the by now well-known ways in which, say, Google biases the results of searches to favour its clients. In any case, the bottom line is that the more actively involved people are in their search for knowledge, the more likely that the underutilized 80 per cent in Pareto's law will receive needed exposure (Brynjolfsson and McAfee 2014: chap. 10).

The Challenge of Academic Rentiership and Its Interdisciplinary Antidote

When Pasteur famously claimed that discovery favours the prepared mind, he was alluding to a mind that was uninhibited by academic prejudice and hence open to a broader range of reality – including that 80 per cent that is already represented in the academic literature but which routinely gets overlooked. That academia might be a source of such prejudice goes back to Francis Bacon's critique of the medieval scholastics as purveyors of 'idols of the mind'. The history of the modern university, starting with Wilhelm von Humboldt's rectorship at the University of Berlin in the early nineteenth century, can be understood as a relatively successful attempt to recover from Bacon's spin on academics as inhibitors rather than promoters of intellectual enlightenment.

Here it is worth recalling that in the two centuries that separated Bacon and Humboldt – the seventeenth and eighteenth centuries – the university was still largely a glorified professional school dedicated to reproducing society's governing elites: law, medicine and the clergy. These professions were, in the first instance, the bastions of continuity against any forces of change. Humboldt exchanged the university's inertial image for a more dynamic and progressive one, in which cutting-edge research would be made part of classroom instruction, so as to ensure that the future leaders in the student body would not mindlessly reproduce the past.

Nevertheless, I say 'spin' because we know now – and Bacon probably knew back then – that not all medieval scholastics fit his influential stereotype. In particular, those who were members of the Franciscan order (e.g. Robert Grosseteste, Roger Bacon, Bonaventure, Duns Scotus, William of Ockham) pioneered both the turn of mind and the turn to experiment which Bacon himself championed and which presaged many of modernity's signature 'progressive' attitudes (Fuller 2015: chap. 2). Be that as it may, Bacon stereotyped the original academics as relying on only one of the two books through which God communicated with humanity: the Bible – but not Nature. And the stereotype did capture how scholasticism was generally understood. Perhaps most notoriously, the mere contradiction of biblical authorities set Bacon's contemporary Galileo at odds against the Catholic Church.

Nowadays economists would diagnose the increasingly esoteric trail of scholastic biblical commentary as symptomatic of a 'path-dependent' and 'rent-seeking' academic culture, whereby once you have discovered a source of truth (in this case, the Bible), you make it difficult for anyone else to seek truth unless they follow your path, if not make specific contact with you – which is the original economic model of 'rent', namely, a toll on the only paved road to your chosen destination. It also provided the context for launching the original attacks on technical language ('jargon') as obscuring rather than illuminating thought, and contributed to Jakob Burckhardt's coinage of 'Dark Ages' to characterize the state of European intellectual life prior to what he called the 'Renaissance'.

In the eighteenth century, Bacon's jaundiced perspective on academic life was adopted with gusto, resulting in several schemes to reorganize academic knowledge in the name of human emancipation from established authority. In particular, much play was made of Bacon's own proposal to base the division of disciplinary labour on our mental faculties – albeit all done while neuroscience was still in its infancy – in order to maximize their spontaneous synergies (Darnton 1984: chap. 5). Because Bacon and his latter-day followers remained convinced of humanity's divine lineage, they were confident that only the blinkered nature of other humans – namely, the clerics who controlled the universities – stood in the way of an indefinitely well-informed and prosperous future.

After all, even though discoveries are advertised as instances of nature revealing itself anew, it is usually not too difficult to find one or more precedents for the discovery in neglected writings from the past. Admittedly, each of these past accounts may be partial, but if the right combination of them had been made, then the innovation could have been inferred. This may suffice as an operational definition of a 'prepared mind' for scientific discovery, in Pasteur's sense. In that case, the 'discovery' is simply a high-tech version of Plato's *anamnesis*, whereby, say, an experimental outcome prompts us to recall what we already implicitly know, thereby enabling us to make of maximum use of 'undiscovered public knowledge'.

During the Enlightenment, the idea that only other humans stand in the way of overall human progress resonated more widely across society, often with revolutionary consequences. In particular, it led to the transition from a capitalist to a socialist mentality with regard to the modernization of political economy. The operative term was 'organization'. As the so-called 'utopian socialist' Count Henri de Saint-Simon (1760–1825) had realized before Marx, capitalism's world-historic job would be left only half-done if it merely overturned the feudal privileges that had restricted free trade. In addition, production itself had to be organized by 'captains of industry' capable of assembling the talents of humanity to result in a whole greater than the sum of its constitutive individuals.

In that case, the enemy is not simple ignorance but *habit*, understood as a kind of second-order ignorance that is reflected in default patterns of behaviour. Habit collectivized is 'tradition', the persistence of which reflects a failure to see alternative ways of combining people's knowledge for more efficient and powerful results – what in terms of this book would be regarded as an underutilization of 'modal power'. This involves not only allowing people to try their luck in the marketplace but also, and more importantly, scientifically empowering the captains of industry to discover people's hidden talents and provide those people the opportunities to exercise those talents – what in the twentieth century came to be called 'human capital' (Fuller and Lipinska 2014: chap. 3). In the next section, we discuss this general sensibility as the 'military-industrial will to knowledge'.

A useful way to understand Saint-Simon's organizational crusade, which inspired the two leading self-avowed 'progressive' movements of the modern era, positivism (through his secretary Auguste Comte) and Marxism, is as pushing the capitalist hostility to rent as a form of wealth creation from land-ownership to self-ownership. Thus, your unused potential as a human being is just as reprehensible as the unused land that you own. Of course, the two situations are not diagnosed or remedied in the same way – but they both constitute a moral blot that society is obliged to clear up. This sense of normative parity between land ownership and self-ownership began with Jean-Jacques Rousseau's attempt to universalize the legalistic understanding of 'ownership' that became the norm in early modern Europe. Its intellectual legacy can be traced through the tortuous history of the concept of alienation, from which the modern understanding of the self/other distinction emerged.

This entire line of thought comes to a head in our day with the rent-seeking tendencies of academic disciplines, which exist in an ontological zone between land- and self- ownership. In this context, peer-review processes do not simply validate proposed knowledge claims, but more importantly create an entitlement to collective ownership on the part of the academic discipline that conducted the validation. As a result, a burden is placed on subsequent researchers to credit such validated knowledge to get one's own knowledge claims validated. This use of the validation process to generate collective intellectual property rights, as branded by the validating academic journal, can be either seen as adding value to the original knowledge claim (i.e. 'asset-building') or simply capitalizing on the monopoly condition under which validators receive the knowledge claim (i.e. 'rent-seeking'). Kean Birch (2017) has led the way in querying this distinction.

That this distinction in motivating the validation process is not normally observed is reflected in capitalist scepticism to market regulators about what, say, ensures safety given the prospect of such policies to impede competition. Regulators claim to be maintaining quality control for goods and services in circulation, but the standard is often set higher than the market players themselves appear to be willing to tolerate, thereby restricting their scope for interaction. To be sure, market regulators claim to speak on behalf of third parties who might be adversely affected by completely free trade (what economists call 'negative externalities'). But these third parties are rarely themselves directly consulted; rather, they are simply presumed by the regulators to be vulnerable. Thus, a latter-day paternalism masks the brute judgement that people are incapable of deciding for themselves the level of risk that they are willing to undertake in uncertain situations. The 'proactionary principle', discussed in the final chapter, was designed in part to cut through this hypocrisy.

In any case, the dynamic just described reveals liberalism's more general difficulty in passing the first hurdle of acceptance to treat people as truly free agents capable of self-legislating. Not surprisingly, perhaps, the precautionary principle, to be discussed more fully in the final chapter of this book, has been an attempt to reinscribe paternalism at a global level by persuading people that it is more rational for them to impede than commit action without having passed through presumably more knowledgeable intermediaries, what Latour (1987) originally popularized in the STS community as an 'obligatory passage point'. (Not surprisingly, over the past 30 years Latour's then-incipient conservatism has become much more full-blown.)

Against this forbidding backdrop, *interdisciplinarity* has been the preferred academic route to remedy 'epistemic rent-seeking', whereby disciplines become increasingly proprietary in their relationship to organized enquiry. The natural opponent of the epistemic rent-seeker is what the sociologist Randall Collins (1979) has called the 'credential libertarian' who sees disciplinarians as George Bernard Shaw famously saw experts more generally, namely, as a conspiracy against the public interest. The advent of the Internet has launched a new and robust wave of credential libertarianism, as we are now always only a few keystrokes away from finding challenges and alternatives to expert opinion on virtually any topic.

As we have seen, a discipline is 'proprietary' in this negative sense if it can compel enquirers to acknowledge its ownership of a field of enquiry, regardless of the discipline's actual relevance to the enquirers' epistemic ends, a 'bait and switch' that results in 'cognitive authoritarianism'. Intuitively put, to know the economy, you must either study economics or defer to an economist. The same is true for knowledge of the physical world (go to physics), life (go to biology), health (go to medicine), (go to sociology) and so on. This 'rent' may take the form of requiring that the enquirers undergo specific discipline-based training or cite authors in the epistemic rentier's field. If organized enquiry is a kind of intellectual journey, then disciplines impose tolls along the way, perhaps for no reason other than their having made a similar journey first.

In short, epistemic rent-seeking is Kuhnian 'normal science' with a clear business plan. This occurs when a distinctive theoretical and methodological framework – or 'paradigm' – with proven effectiveness is extended indefinitely, thereby crowding out other approaches and the distinctive sorts of questions that they potentially address. For example, after the early striking successes of Newtonian mechanics, all of material reality was presumed to be covered by the framework, which served to drive many tricky issues relating to the nature of life and mind into the margins of physics, which resulted in the disciplines of biology and psychology (Fuller 1988: chap. 9; Prigogine and Stengers 1984).

But these disciplines have also undergone the same 'normalizing' process. The endgame of this vision is that domains of phenomena come to be owned by disciplines, access to which involve high entry costs, not only in terms of specialized training but also the human and material resources required to bring a research project to fruition. At the same time, and perhaps due to this resource intensiveness, there is little incentive for significant disciplinary reorientation. Indeed, according to Kuhn, a discipline needs to be on the verge of self-destruction – the 'crisis' that precipitates a 'revolution' – before its paradigm is properly replaced.

Thus, disciplines are 'path-dependent' entities, whose very success in following a particular path becomes a strong attractor for other fields of enquiry. The proven success of Newtonian mechanics became a template for other disciplines to follow. Considerable philosophical midwifery has been involved in the process, to be sure, most heroically in the twentieth century by the logical positivist movement. Here we see epistemic rent-seeking at its subtlest, especially with regard to the social sciences. After all, Newtonian mechanics in its nineteenth-century heyday – concerned as it was with light, electricity and magnetism – did not offer any special access to the phenomena studied by the social sciences. On the contrary, as James Clerk Maxwell made patently clear in his 1870 presidential address to the British Association for the Advancement of Science, physicists were the ones who had to adopt styles of probabilistic reasoning already being championed in the field that Adolphe Quetelet had dubbed 'social statistics' (Porter 1986).

By following Maxwell's advice, the next generation of physicists witnessed revolutions, first in thermodynamics and then quantum mechanics. Nevertheless, in Maxwell's day, physics as a discipline was already seen to have found a formula for epistemic success, effectively establishing the 'gold standard' of scientific status, in terms of which aspiring candidates to the title 'science' could be judged – not least, the social sciences themselves. Notwithstanding the robust presence of a more literary, humanistic tradition in the social sciences throughout the twentieth century, the need to approximate physics-based standards of epistemic achievement took its toll, in the form of marginalizing research that did not easily fit into the physics mould. Indeed, when people speak of 'interdisciplinary' research in the social sciences today, they often mean nothing more than bringing its 'interpretivist' and 'positivist' sides back together in a manner that would not be out of place on the pages of, say, Marx, Weber or Durkheim, all of whom wrote more than a century ago.

Logical positivism plays a Janus-faced role in the history of epistemic rent-seeking. I have already suggested that it turned physics into the high-rent district of organized enquiry. Yet, at the same time, the positivists demanded that other disciplines explain why they required theories and methods that differ from those of physics. In this respect, logical positivism resembles the sort of liberal imperialism promoted in Victorian Britain. Both were officially 'free trade' doctrines designed to promote relatively frictionless transactions in ideas and goods, respectively. But equally, both assumed a privileged position from which to espouse the free trade doctrine. In the case of the positivists, privilege was conferred on mathematics, be it symbolic logic or statistical representation.

When I was a student, this feature of logical positivism was expressed as a distinction between the 'context of discovery' and the 'context of justification'. Science as an institution converts the idiosyncratic origins of discoveries into knowledge claims that anyone in principle can justify for themselves simply by examining the evidence and reasoning offered for a particular knowledge claim. In this way, individual insights come to be incorporated into a collective body of enquiry, which in turn empowers humanity as a whole. Thus, while a particular truth may have been discovered in a very particular way, the task of science is to show that it could have been uncovered under a variety of circumstances, provided the necessary evidence and reasoning (Fuller 2000a: chap. 6). In short, when science is doing its job, path dependency is broken.

It is easy to see how this positivist principle could sound the death knell to epistemic rent-seeking. The positivists themselves – much in the spirit of past imperialists and today's globalizationists – saw the removal of trade barriers as leading to greater integration and interdependency. Interdisciplinarity would be effectively fostered through anti-disciplinarity, at least insofar as disciplines would need to translate their specific jargons into a common lingua franca of intellectual exchange. Yet, this is not quite what happened. Just as in the economic case, the already existing power asymmetries between the disciplines played themselves out in this 'free trade zone'. While many disciplines became physics-friendly, non-physics-friendly modes of enquiry were consigned still further into the intellectual backwaters. Mathematics constituted a hidden barrier to free trade in this context. The result is that epistemic rent-seeking and its proposed positivist remedy nowadays coexist without resolution, arguably the worst of both worlds. Again, this is not so different from the world's political economy – and the phenomenon is especially evident in the social sciences.

On the one hand, there are the economists and psychologists capable of using mathematics to arrive at striking conceptual insights that their natural science peers can easily understand and appreciate. These can then lead to interdisciplinary alliances, which are on display, say, at the Santa Fe Institute. On the other hand, less mathematically inclined enquirers can fall back on their 'indigenous' disciplinary knowledge bases. This helps explain the increase in what might be called 'referential entrenchment' – that is, high citations to a well-bounded body of texts. In that case, one needs to master the local jargon before making any intellectual headway: very high rent but in a relatively small, gated community. Much of contemporary academic 'postmodernism' can be understood this way, and not only by its detractors. It can be painful to watch colleagues who start by discussing, say, the sources of inequality descending into a discussion of the relative merits of, say, Michel Foucault's and Pierre Bourdieu's views on the matter.

The Cult of Success and the Military-Industrial Will to Knowledge

The state of mind required for the acquisition of knowledge started to undergo a significant transformation in the Christian era, which continued into secular modernity. The Greeks believed that 'knowledge', in the sense of a systematic understanding of how things hang together, was available to all educated people with sufficient leisure at their disposal. The relevant mental states included Platonic contemplation of the cosmos and the receptive observation of nature promoted by Aristotle. Notably absent was the idea of a *journey*, let alone one whose destination is at best partially understood in advance and may not even be reached by those who start the journey. Nevertheless, both in terms of Christian eschatology (the theory of human destiny) and modern secular conceptions of progress, such a journey was necessary because humans exist in a state well below their full potential for being in the world, be that due to original sin or some entrenched habits of mind. This then became part of the collective psychology required for justifying science as an intergenerational social enterprise of indefinite duration (Passmore 1970: chaps. 10–12; Fuller 2015: chap. 3). The military-industrial 'will to knowledge' puts a premium – perhaps more so than academics – on reaching a satisfactory conclusion to the epistemic journey.

That academics are inclined to stress the virtues of the journey over its destination may be dubbed the 'Curse of Kant' – the academic reluctance if not incapacity to specify an end to enquiry that would count as a meaningful conclusion to those engaged in it. Instead, academics tend to provide 'regulative ideals', à la Kant himself, which specify little more than an orienting posture to enquiry or, à la logician Alfred Tarski, 'truth conventions', which identify conditions which need to be met for 'truth' (the name given to the end of enquiry), while saying nothing about how those conditions might be met. Missing from all this is a strong sense of *success*, which is central to the military-industrial will to knowledge.

Success implies that a journey has ended in a way which vindicates its having been taken. In this concrete sense, the end will have justified the means. Here it is worth recalling that 'the ends justifies the means' was originally a theological slogan to justify how all the evil in the world contributes to the realization of divine creation. Kant was a sworn enemy of this way of thinking, but less on ethical than on epistemological grounds: if God's existence cannot be proven, then trying to think like God is idle – and potentially dangerous – speculation which ignores something that is already known, namely, human finitude.

In contrast, the military-industrial will to knowledge adopts a standpoint that was openly defended two generations after Kant by such left-wing followers of G. W. F. Hegel as Ludwig Feuerbach and Marx. It takes 'God' to be simply the name of the successful conclusion of the project of human self-realization. In that case, any obstacles in the way to getting 'what we want' (however defined) is taken more in the spirit of a challenge to be overcome than a sign of limits that should not be exceeded. This line of thinking, which had explosive political consequences in its day (i.e. the 1848 liberal revolutions), was encoded in philosophy's DNA by the German-trained Scottish philosopher who introduced 'epistemology' into English, James Ferrier (Fuller 2015: 34).

As if in anticipation of Wittgenstein's *Tractatus*, Ferrier argued that if the world can be divided exhaustively into what is 'known' and what is 'unknown', it follows that everything is knowable, which means that the very idea of 'unknowable' (e.g. Kant's *noumenon*) is nonsensical. The 'unknown' is simply the 'yet to be known'. (This applies equally to the 'unknown unknowns', more about which below.) In this way, a much more concrete and explicit sense of the ends of enquiry – a successful conclusion to a strategy that one might design and execute – is licensed, which in turn has served to propel the military-industrial will to knowledge over the past 150 years.

However, the 'military' and the 'industrial' sides of this intellectual complex have paradigmatically different ways of interpreting 'success'. The difference is illustrated in two senses in which we might achieve the 'truth', understood as the concrete ends of enquiry. We might mean either something to which one gets closer or something that grows over time, an 'intensive' or an 'extensive' magnitude: the former measured, the latter counted. Thus, one speaks of either 'approximating the truth' or 'accumulating truths'. That is the difference between 'military' and 'industrial' approaches, respectively (Fuller 1997: chap. 4; Fuller 2002: conclusion).

The former is mainly about how to make up the distance from a goal, the latter about how to expand so as to incorporate – or crowd out – all other competitors, which is itself constitutive of the goal. Thus, two conceptions of success loom: *victory* in war and *monopoly* in commerce, the one material-ized in the correspondence theory of truth, the other in the coherence theory of truth. Capitalism plays a distinctive role in this dialectic, with its fixation of *productivity*, which is about not only producing more stuff (industrial) but also producing it more efficiently, which may mean abandoning or destroy-ing past practices (military). It was just this dual feature of capitalism that led nineteenth-century socialists – not least Marx – to envisage a less labour-intensive future, courtesy of new technology, which meant that most people would live either in leisure or poverty, depending on what Marx called the 'social relations of production'. In Pareto's terms, the industrial side of capital-ist productivity belongs to the lions, the military side to the foxes. Now let us apply this turn of thought to the production of scientific knowledge.

The military-industrial will to knowledge is clearly *anti-disciplinary*, but how is it *interdisciplinary*? It is anti-disciplinary by virtue of its denial that there is something 'natural' about the origin and maturation of disciplines. Thus, the military side would hasten knowledge production, capitalizing on what-ever urgency might focus minds on overcoming a common foe, whereas the industrial side would scale up knowledge production, enabling it to escape the laboratory and permeate the lifeworld. From this dual standpoint, the organic 'no science before its time' approach to enquiry championed in Kuhn (1970) looks like a mystification of the history of science. Indeed, rather than tend-ing to disciplines as protected species in an academic ecology, the military-industrial will to knowledge treats them in the manner of plant and animal husbandry, open to the mixing and matching of knowledge from multiple dis-ciplines to improve the human condition. The Franco-Prussian War of 1870–71 had demonstrated how the exigencies of war can incentivize advanced industrial economies to streamline production processes, distribution net-works and accelerated innovation. The US-based Carnegie Endowment for International Peace, one of the world's most influential think tanks, funded Vienna Circle founder Otto Neurath and British economist John Maynard Keynes to explore the matter in light of the Balkan Wars and the First World War, respectively.

Although by disposition Neurath was a socialist and Keynes a liberal, both recognized that the perceived threat of war strengthened the state's hand in providing incentives for industry to alter its default patterns of behaviour in more socially beneficial ways – not least the retention of novel interdisciplin-ary working arrangements forged in wartime. This is what the mastermind of the Prussian victory in the Franco-Prussian War, Baron Helmuth von Moltke, had called the state of 'permanent emergency', about whom more below. In the middle third of the twentieth century, economics provided the main disciplinary arena for concretizing this vision. The venue was the Cowles

Commission, named for the owners of the *Chicago Tribune*, which by the 1950s had successfully promoted econometric modelling – the prototype for today's computer simulations – across the ideological spectrum of economics, not only the Keynesians but also the Soviets (via Oskar Lange's market socialism) and even neo-liberal capitalists (via Milton Friedman's monetarism). At the time, this development was seen as having moved economics decisively away from its classic 'free market' base, the sense in which Hayek might be seen as the natural heir of Adam Smith, since Cowles clearly took a dirigiste approach to the economy – that is, as something that might be controlled as a machine, a cyborg version of Homo economicus (Mirowski 2002: chap. 5).

At stake here was whether human creativity, the source of the entrepreneurial spirit, could be simulated in some systematic understanding of economic life. Among Cowles's biggest opponents were Adam Smith's self-appointed US descendants, most notably Frank Knight, who started the Chicago School of Economics. Knight rejected Cowles's top-down approach in favour of something 'spontaneously generated' but also mysterious; hence, Knight influentially distinguished between states of ignorance due to 'uncertainty' proper (i.e. where no probability can be assigned because of the event's unprecedented nature) and to 'risk' (i.e. where probability can be assigned because the event has precedent). The former was the realm of the entrepreneur and the latter that of the manager. For this reason, Knight held that the use of mathematics in economics would always be restricted to routine ('manageable') economic behaviour but not the creative part, a claim similar to one often made about the limits of experimental psychology in understanding human creativity.

However, Knight failed to anticipate how mathematical models would radically transform thinking about uncertainty. It amounted to the management of 'unknown unknowns', as US defence secretary Donald Rumsfeld memorably put it during the Iraq war, by converting them into 'knowables' in a simulated universe defined by a set of simultaneous mathematical equations, on the basis of which various projections could be made.

Here Rumsfeld was offering a high-tech re-enactment of the line of reasoning which had led Ferrier to invent epistemology 150 years earlier. Anything that counts as an 'unknown unknown' is already knowable, which is to say, it exists – and, to recall Willard Van Orman Quine's quip, 'to be is to be the value of a bound variable' (in, say, a set of simultaneous equations). In other words, the 'unknown unknown' must satisfy certain parameters, which may be operationalized in various ways. Thus, while one might not know exact values, one would know 'the margin of error'. More generally, be it understood as happening in a 'planned' or 'blind' fashion, the economy might be modelled by algorithms that coordinate independent streams of input data into a concerted response. Indeed, on the basis of such systems-level thinking, one might try to achieve 'planned' outcomes by 'blind' means, or vice versa – which basically captures the political-economic space defined by market socialist and social democratic modes of provision (Fuller 2016a: chap. 4). In principle, everything from the flow of money to the allocation of resources could be subject to this treatment.

Thus, the great Cold War interdisciplinary project, *cybernetics*, was born (Mirowski 2002: chap. 2). At the height of the Cold War, Massachusetts Institute of Technology (MIT) political scientist Karl Deutsch (1963) conceptualized the state as society's cybernetic brain, a metaphor that the UK management consultant Stafford Beer tried to render literal as the mastermind of Chile's ill-fated attempt to implement cybernetic socialism ('Project Cybersyn') via a mainframe computer which was located in a room destroyed during Augusto Pinochet's 1973 coup (Pickering 2010: chap. 6). In any case, many of devotees of this approach, not least Rumsfeld, made successful careers by applying the template. For better or worse, the exemplar of applied cybernetics (aka operations research) was Robert McNamara, whose early conversion to systems-level thinking at the Harvard Business School enabled him to run Second World War bombing raids over Japan, turn around the fortunes of the Ford Motor Company, direct US military operations in Vietnam and finally preside over the World Bank's push into development aid (Fuller 2000b: 183).

The most radical and ultimately persuasive formulation of the military-industrial will to knowledge came from the man who led the Prussian troops to victory against France in 1871. According to Moltke, a society should regard itself as always in a state of 'permanent emergency', which amounts to thinking in terms of who or what might *next* threaten its very survival (Fuller 2000b: 105–9). This alerts the society to the need to reassert its existence in the face of ever-changing circumstances. In short, peacetime is when you learn how to fight the next war. Thus, the image of society as organism and as system became fused in society's never-ending struggle to define the boundary between itself and the external environment. The significance of this fused image should not be underestimated. (It will be relevant to 'superforecasting' in the final chapter.) After all, 'system' had been normally seen as a term in logic, the formal relations of parts to wholes, but now it was enmeshed with a substantive biological imperative whereby only those who advance (or grow) survive (or live). As we shall see, this epistemologically deep albeit paranoid vision came to fruition during the Cold War.

The organicist conception of society played into the nineteenth-century conception of the state as guardian of a 'nation', which is to say, a political entity grounded in the existence of a native population (to which non-native citizens may or may not be added). It also played into such contemporaneous developments as Claude Bernard's 'experimental' definition of death as an organism's failure to maintain a strong distinction between itself and its environment (i.e. death as blending) and Leon Walras's general equilibrium model of the economy as a self-sustaining mechanism. While Bernard's definition of death came to be seen as a special application of the principle of entropy in thermodynamics, Walras's model scaled up into a vision of society as a set of interlocking markets, which were kept in harmony by a regulatory state. Such a world view was championed by Walras's younger contemporary Pareto, who in turn was translated and promoted in the United States by Harvard biochemist Lawrence Henderson. Henderson's colleague, the physiologist Walter Cannon, coined the term 'homoeostasis' to capture this line of thought in its maximum generality. It was picked up by the sociologist Talcott Parsons, who aspired to an interdisciplinary science of 'social relations', and especially the mathematician Norbert Wiener, in whose hands it became the cornerstone of cybernetics (Heims 1991: chap. 8).

For those following in Moltke's footsteps, 'war' is simply the violent version of ongoing conflict that is normally played out in the marketplace. Just as rival producers aim to prevent each other from acquiring a monopoly, nation states regard their rivals as promoting alternative hegemonic regimes, which they then try to stave off. In this respect, 'peace' is simply the sublimation of hostilities posed by the rival universalisms. The logic underwriting this train of thought is not so hard to spot.

But Moltke's idea truly came into its own in the Cold War, when first the laboratory and then the computer emerged as hybrid military-industrial platforms for the playing out of interstate rivalry. Thus, starting with the Soviet launching of Sputnik in 1957 and including the rival US-USSR space missions and the various iterations of the 'arms race', the mere display of techno-scientific strength mattered more than its actual deployment, either in war or commerce. In the end, the United States won the Cold War simply by outspending, effectively bankrupting the USSR. Thereafter many of the innovations that had been capitalized during the Cold War were quickly released into the marketplace, resulting in Silicon Valley and its various global emulators driving the post–Cold War economy. This in turn has opened up new national security concerns, captured nowadays in the various biotech hybrid discourses of 'virus' (Mazzucato 2013).

Moltke had been inspired by Karl von Clausewitz's early nineteenth-century classic, *On War*, which is famous for defining war as 'politics by other means'. But he then took Clausewitz to the next level, arguably perfecting the 'art of war'. Here it is useful to think about war as proceeding along a dialectic. In the first moment, war is a natural extension of everyday conflict with no agreed rules or objectives – that is, until the parties see it as no longer in their interest to fight. Thus, Fabius, the third-century BCE Roman general credited both with Hannibal's defeat in the Punic Wars and the name of the United Kingdom's 'Fabian' socialist movement, simply wore down his stronger opponents by forcing them to fight on his own turf, which then enabled his troops to win by deploying guerrilla tactics. In contrast, warfare in its second moment acquires clear strategic goals and mutually agreed rules of engagement of the sort that makes Clausewitz's definition ring true. This remains the default 'modern' sense of war, in which the making and breaking of treaties between nations provide the official record of geopolitical affairs.

However, Moltke aspired to make war the limiting case of the human condition, which marks the third moment of the dialectic. In true Hegelian fashion, it reintroduces elements of the first Fabian moment. Thus, Clausewitz's well-defined opponent in a structured conflict also provides the environment through which one's own more existential objectives are realized. In this respect, the opponent is internalized as a means to one's own ends. Put bluntly: we need opponents to fully realize who we are and hence what is worth defending and promoting. The Cold War's self-consciously Manichaean struggle between capitalism and socialism epitomizes this equation of life and war, without any obvious end in sight. In concrete terms, it marked the advent of the military as the most proactionary agency of the state in terms of its approach to science and technology.

This expanded ambition to be a general science of strategic control appeared in a shift in military procurement policies from what was needed to fight the next version of the Second World War in the foreseeable future to what was needed to fight a genuinely novel 'Third World War' in the indefinite future. In this context, the UK political economist Mary Kaldor (1982) has written of the rise of a 'baroque arsenal' that is preoccupied with rapid response and action at a distance as ends in themselves, regardless of specific short-term advantage. The relevance of this point to interdisciplinarity is that whenever the exemplar of this approach – the US Defence Department Advanced Research Projects Agency (DARPA) – is stereotyped as harbouring 'mad scientists', their 'madness' typically relates to the scientists' refusal to accept the normal spatio-temporal parameters of discipline-based research (Belfiore 2009).

The Corporation as the Hub of the Military-Industrial Will to Knowledge

To appreciate the issues at stake in this section, we need to take a brief detour through corporate history. Before the nineteenth century, the legal category 'corporation' as an agency independent of the state with a distinct legal personality was reserved for civic, academic and religious organizations. In this respect, the Royal Society was more of a 'corporation' than the joint-stock companies that we now take to have been the original vehicles of capitalist expansion in seventeenth- and eighteenth-century Britain. Those companies ultimately depended on the favour of the monarch, who functioned as the principal shareholder, typically taking a personal interest in matters. However, the second half of the nineteenth century witnessed a wave of economic liberalization that allowed for the self-perpetuation of businesses on the corporate model, including limited liability to owners and rights to expansion, especially when seen to be in the public interest. Thus, 'corporations' in this modern sense were permitted to minimize transaction costs by acquiring product-relevant supply chains and distribution channels. But just as importantly, corporations were allowed to throw open ownership to the marketplace, specifically the stock market, whereby the non-controlling 'shareholder' became the norm.

However, as Adolf Berle and Gardner Means (1932) famously showed in *The Modern Corporation and Private Property*, the net result of these shifts in corporate law was to drive a wedge between the ownership and control, and in the process deconstruct the concept of property. The market's democratization of owners as 'shareholders' served to diminish corporate accountability, leaving it to a class of professional 'managers' whose jobs depended on securing ever larger yields from shareholder investments – by whatever means. Relevant for our purposes is that the risky decisions taken by this new managerial class were typically informed and legitimized by interdisciplinary researchers employed by foundations that were financed by these very corporations. Nevertheless, at the time, which coincided with the rise of communism and fascism, the point remained obscured. Instead, *The Modern Corporation and Private Property* was read as having demonstrated the complementarity of mass democracy and authoritarianism, insofar as shareholders did not seem to care what managers did as long as dividends increased. And by the time the managers failed, it was often too late to change course without significant losses, which historically culminated in the Great Depression. It was against this backdrop that US President Franklin Roosevelt extended the state's role as public interest regulator of the business world. FDR turned to such interdisciplinary researchers as Berle, himself a corporate lawyer, to constitute a 'Brain Trust' that would draft legislation designed to insure against any future corporate risk-taking.

The emergence of corporate foundations – Rockefeller, Carnegie, Ford, Sloan, and so on – reflects a unique confluence of forces in early twentieth-century US political economy. At the intellectual level, there was the recognizably Baconian desire to unleash practical knowledge from its scholastic shackles, which extended to the opening up of research to areas of social life – most notably the workplace and the family – which academics had pre-viously ignored. Why the neglect? The most obvious reason was that these settings lacked the elaborated textual traces with which academics were most comfortable. Thus, until the early twentieth century, 'history' generally meant the study of the speeches, memoirs, treatises and treaties of major political players. Given that the quintessentially academic field of 'cultural studies' rou-tinely valorizes ordinary vis-à-vis elite lives nowadays, it is easy to forget that with a few notable exceptions (e.g. Friedrich Engels's early ethnographies of British working-class lives) most of the research into non-elites was originally commissioned by the state or industry. To be sure, the motivations for these commissions were often self-serving, but at least these powerful social actors recognized that their legitimacy depended on the well-being of the general population – an insight which did not naturally occur to academics.

Beyond this intellectual impetus, which also characterized the roughly con-temporary rise of corporate foundations in Germany and elsewhere, there was the political need for American industrialists to adapt to the normative horizons of the emerging Progressive movement, whose default view of big business was one of tax-dodging monopolists, the ultimate rogues of public life. Thus, John D. Rockefeller, threatened with a big tax bill from oil revenues, created the Rockefeller Foundation, the aim of which would be to finance research that aimed to increase worker productivity, presumably as a means to recover the lost tax money. Jane Mayer (2016) has interestingly called this 'weaponized philanthropy'.

Nevertheless, industrialists like Rockefeller shared the Progressives' world-historic sense of American nationalism. Thus, they positioned themselves to complement the emerging proactionary state championed by Theodore Roosevelt and Woodrow Wilson. Specifically, the foundations were key in resourcing foreign scholars whose work was disrupted by the two world wars, as well as providing the main financial support for research relating to the improvement of worker productivity and, at a still deeper level, health and education. The overall strategy worked, as the United States became a research leader of global standing, long before the establishment of a 'National Science Foundation' in 1950. Indeed, the historian Charles Beard could already observe in 1936 that the modus operandi of foundations was to manufacture and repair 'cultural lag', as they generated disruptive innova-tions in the home and in the workplace which provided a pretext for them to collaborate with the state to move the masses to ever newer – and presumably higher – senses of 'normal' (Kevles 1992: 205). In effect, the foundations had become the creative destroyers of the American way of life.

The response of university leaders to this agenda was mixed. Among those who fell out of love with the foundations' creative destructive impulses was Robert Maynard Hutchins. As Dean of Yale Law School in the 1920s, Hutchins invited the Rockefeller Foundation to establish an interdisciplinary social science institute to study and remedy liabilities in legal judgement, thereby seeding the modernizing movement known as 'legal realism'. But as President of the University of Chicago in the 1930s, Hutchins abruptly arrested the development of the social sciences (and opened the door to the more 'humanistic' approaches associated with Leo Strauss and Hayek) by refusing Rockefeller money, which he believed was responsible for the market volatility that resulted in the Great Depression. On this basis, Hutchins became a convert to 'perennial philosophy' and natural law theory in the neo-Aristotelian mould (Ross 1991: 400–3). In striking contrast, upon becoming Harvard's President in 1933, the chemist James Bryant Conant made it much easier for the foundations to access Harvard faculty and resources. As the first scientist to run an Ivy League institution, Conant was more comfortable with the socially disruptive character of scientific innovation, which academics were ideally positioned to study, manage and capitalize on. Here he had been inspired and promoted by the Harvard biochemist Henderson, mentioned in the previous section, who worked with the industrial sociologist Elton Mayo on the famous Rockefeller-funded 'Hawthorne' studies on the environmental factors informing worker productivity (cf. Fuller 2000b: chaps. 3–4; Isaac 2011: chap. 2).

The largely symbiotic relationship between corporate foundations and big government in the United States in the first half of the twentieth century amounted to an implicit division of labour over research funding. At first, the state treated science and technology policy as an extension of domestic development policy, largely on the model of the land-grant universities established in rural regions in the nineteenth century. The focus was on applied research. Indeed, when the prospect of a permanent 'National Science Foundation' was debated in the years immediately following the Second World War, MIT vice president Vannevar Bush had to coin the phrase 'basic research' to capture another basis – the one that ended up prevailing – on which the state might wish to fund science (Fuller 2000b: 151). In contrast, the foundations always knew about basic research, which they practiced in the spirit of Stokes (1997), discussed at the start of this chapter. From their standpoint, a field like 'molecular biology' (a Rockefeller coinage) was about improving the capital stock from which better products might flow in the future. But here 'capital stock' refers to not only better crops and livestock but also better humans. Thus, while a discipline-driven view of knowledge might view 'eugenics' as the (mis) application of genetic science, the foundations saw eugenics and other broadly 'social' sciences in 'basic research' terms that aimed to extend human potential indefinitely (Kay 1993).

This conception of basic research meant that the foundations had to think creatively about how to navigate their investments through the twentieth century's political-economic volatility, much of it arguably of its own creation. For example, in the wake of the Great Depression, as director of natural science research, Warren Weaver – now remembered for the Shannon-Weaver entropy-based communication theory – steered the Rockefeller Foundation away from providing seed money and towards funding extant discipline-based research that could use the money to develop synergies in interdisciplinary projects (Kohler 1994: part 3). This was the spirit in which Weaver first identified 'molecular biology' as a field of interest for the foundation in the 1930s, which incentivized physicists and chemists to migrate to biology, eventuating 20 years later in the discovery of DNA's double-helix structure. Interestingly, the US National Science Foundation in the first decade of the twenty-first century adopted a similar strategy to promote a broadly 'transhumanist' agenda to bring together nano-, bio-, info- and cognosciences in common projects aimed at 'enhancing human performance' (Fuller 2011: chap. 3). But generally speaking, after the creation of the NSF in 1950, the foundations took on riskier interdisciplinary research assignments – such as radiation-based medicine and mechanical models of the brain (aka neuroscience) – which were unlikely to survive the discipline-based peer-review processes instituted at the NSF (Kevles 1992: 209, 221).

Finally, the long-term effect of the foundations on the corporate form of business itself is worth pondering. The lifework of Alfred Chandler, America's premier business historian, can be understood as having been about the disintegration of the modern corporation through a pincer attack from natural and social scientists in the twentieth century. Chandler himself did not put matters this way, but it captures the dynamic described in Chandler (1962), which recounts how the corporation moved from being organized in *functional* to *divisional* terms, the former characterized by mass production and the latter by market specialization: 'Fordism' and 'post-Fordism', as cultural sociologists put it. (Or, in more Biblical terms, 'Leviathan' and 'Behemoth'.) The research legitimizing the former was natural science-led, the latter social science-led. Frederick Winslow Taylor and Abraham Maslow are the respective patron saints of these approaches. On the one hand, you might strive to strive to produce something so inexpensive yet useful that you conquer the market through mass acceptance. On the other hand, you might strive to customize what you already produce to distinct markets, even though that may mean investing more in learning about your customer base than in physically changing your product (Stinchcombe 1990). Yet, this bipolar strategy has arguably led to the business world's version of 'imperial overreach', whereby the corporation's identity (or 'brand') becomes compromised by trying to be all things to all people.

Conclusion: The Cautionary Tale of Fritz Haber and Larger Lessons for Interdisciplinarity

A theme that has run throughout our account of the military-industrial route to interdisciplinarity is the Promethean potential of 'undiscovered public knowledge', those products of epistemic associations that stray beyond the rent-extracting path dependencies of established paradigms. A hundred years ago the problem was felt in terms of the knowledge outside the academy that resulted from this implied neglect. Henry Etzkowitz (2002) credits Vannevar Bush, when working at the MIT electrical engineering department in the 1920s, with recommending that academics regularly consult with business to get new ideas and even form what are now called 'start-up' firms. A Bush student, Frederick Terman, took this inspiration to the US West Coast, where as Stanford University Provost in the 1950s, flush with federal funds, he built the academic hub for today's Silicon Valley. While a question mark remains over the ultimate economic efficacy of the information technology revolution unleashed from Silicon Valley since the end of the Cold War, so far it has been generally seen as a 'good thing'. But as some of the more hyperbolic claims relating to 'artificial intelligence' come closer to realization, storm clouds may be brewing – and here history may provide an instructive precedent.

A key transition in the evolution of interdisciplinarity came with the adoption of what is nowadays called the 'triple-helix' model of state-industry-university relations, the signature institutional formation of 'mode 2' or 'post-academic' knowledge production (Gibbons et al. 1994). When the origins of this model are traced to the post–Cold War era, marking the end of state-protected academic enterprises and the onset of a more 'postmodern' and 'neo-liberal' sensibility, it is easy to see the transition as hostile to classic academic ideals of research autonomy. However, if one goes back to the model's true origins in Germany's Kaiser Wilhelm (now Max Planck) Institutes in the early twentieth century, the picture looks rather different. What had been already recognized for a century as academia's centrality to the long-term national interest was extended from crafting future leaders in the classroom to crafting innovative products in the laboratory, as academics were politically permitted a freer hand in dealing with industry (Fuller 2000b: 107–8).

The exemplar of the massive power – for both good and ill – unleashed by the original triple-helix model is Fritz Haber (1868–1934), who for more than two decades worked at the Kaiser Wilhelm Institute for Physical Chemistry. Haber was awarded the 1918 Nobel Prize in Chemistry for discovering the ammonia synthesis process, which is responsible for both the artificial fertilizers that have provided sustenance for billions of people and the chemical weapons that have killed millions of the same people in warfare over the past hundred years. (In the latter case, I mean the explosives he developed from nitric acid, but he also more famously worked on chlorine- and cyanide-based 'poison gas' weapons.) What is now known as the 'Haber-Bosch process' involves harnessing nitrogen from the air – 78 per cent of the Earth's atmosphere is nitrogen – instead of having to rely on material deposits.

Despite the idiosyncratically interdisciplinary character of Haber's work, it had real-world effects, given his easy access to heavy industry's capacities to scale up the production and distribution of innovations. While industry's role in Haber's success is generally noted, starting with his father having been a dye manufacturer, the interdisciplinary character of his thinking tends to be overlooked. Nevertheless, surprisingly for a 'physical chemist', Haber's tripartite doctoral examination reveals someone excellent in philosophy, adequate in chemistry and poor in physics (Charles 2005: 19–20). Two marks of Haber's interdisciplinarity are relatively obvious. The first is the geopolitical significance of maintaining bread as a food staple in the West, amidst growing rice production in the rapidly modernizing East. Haber quickly enabled Germany to become agriculturally self-sufficient, no longer needing to import nitrogen-rich guano from Latin America. This striking success launched the ongoing political discourses over 'food security' and, more generally, 'energy self-sufficiency' (Smil 2001). The second interdisciplinary feature is the conversion of insecticides (largely to protect the new artificially grown crops) to munitions for use against humans, thereby shifting the ontological status of the enemy to that of an animal, which introduced a mode of strategic thinking into warfare that the Nazis would make explicit, not least via 'Zyklon B', the insecticide developed by Haber which was used in the gas chambers of the Holocaust.

However, a third mark of interdisciplinarity was perhaps the deepest and appealed to the Kaiser Wilhelm trustees. It was the idea that ammonia synthesis allowed for the creation of 'bread from the air', in the original marketing phrase for artificial fertilizers. This was basically a secularization of the Biblical 'manna from Heaven', a reference to God's provision of purpose-made sustenance to the Israelites in their 40years of desert wandering from the time of escaping their Egyptian captors to the arrival at the Promised Land (Exodus 16; John 6). The Kaiser Wilhelm Institute's main ideological defender was the German higher education minister, the liberal Protestant theologian Adolf von Harnack, in whose house Haber conducted some of his most influential seminars during the First World War, which Harnack notoriously announced would be won by Germany because of its unique confluence of divine and scientific power (Charles 2005: 118).

If greatness were measured simply in terms of control over the number of lives *both* made and lost across the planet, Haber would have to be considered the single greatest person of the twentieth century, if not all time. Indeed, the non-scientific work of one of Haber's most promising understudies, Michael Polanyi, may be understood as having been written in violent reaction to this prospect, which he diagnosed in terms of Haber's instrumentalist philosophy of science (Fuller 2000b: 139–40). Moreover, Haber-like claims used to be made about the original subatomic physicists, yet interestingly nuclear fission – let alone fusion – has yet to match the level of both benefit and harm wrought by the ammonia synthesis process. Just as we have not had to endure a thermonuclear war, neither have nuclear power plants provided cheap and safe energy to most people. Of course, both could change in the future, perhaps with the help of a Haber-like interdisciplinary genius equipped with artificial intelligence.

Most theoretical discussions of interdisciplinarity treat it as something that is achieved within the academy, with more or less resistance from fellow academics and with more or less pressure from outside the academy. The military-industrial route to interdisciplinarity challenges this starting assumption by denying much of the sovereignty that academics have over the knowledge they produce. 'Sovereignty' here covers not only the uses to which academic knowledge is put but also the very means by which the knowledge is produced. The military and industry have been emboldened in this way for several reasons that have surfaced in this chapter. The main one has been that academic knowledge production tends to be strongly path dependent (aka paradigm driven) and underutilized to the point of being ignored even by other academics. Behind this concern is a faith that the key to human salvation is the full realization of our (god-like) cognitive capacities. Thus, we hold ourselves back by not taking full advantage of all that we know – and can know. Lest one find this faith naive or fanciful, it is worth recalling that those who have promoted the military-industrial will to knowledge have been themselves academically trained. They have experienced first-hand the empowering character of academic knowledge, which they may now find their teachers selling short. In brief, what has been provided here is an account of interdisciplinarity from the standpoint of the successful alumni.

Appendix: Prolegomena to a Deep History of 'Information Overload'

This appendix aims to excavate several senses of 'information overload' which affect all knowledge processes at different levels. The argument begins at the most systemic level, considering information overload as a by-product of the sheer reproduction of knowledge, what I call the 'original position of knowledge'. The invention of the printing press changes the character of this problem, which I illustrate by examining the dual legacy of Newton's *Principia*

Mathematica for the humanities and the sciences, the one focused on 'originality' and the other on 'priority' as overarching epistemic values in dealing with information overload. Legacies of both can be felt in the approaches that authors and readers have taken to texts in the modern era, even outside of the academic fields. Roughly speaking, humanists (notably Rousseau) effectively transcended the problem of information overload by inventing new forms of writing, whereas scientists reinterpreted the problem as a disciplining device to focus their enquiries. The strengths and weaknesses of both strategies are explored. The appendix ends with a consideration of information overload at the individual level, where it is seen as an impediment to decision-making. Here I focus on the role of the decision itself as a creator and displacer of information.

According to historian Ann Blair (2010), the idea of 'information overload' has late medieval roots, related to the proliferation of texts, each reproducing everything the reader needs to know, to see not only the author's contribution but also the author's competence to make said contribution. Of course, this meant that each text said 'everything' somewhat differently, some of which involved unintentional error and some a deviation by design. Thus, creativity is an inevitable consequence of the reproduction of knowledge over time. One might dub this, in a Rawlsian vein, the *original position of knowledge*, the frame of reference that unleashed the first modern revolution in scholarship, associated with the Renaissance, which sought to resolve these intertextual discrepancies by the recovery of 'original' documents which carried normative force. The literary critic Harold Bloom (1973) artfully repurposed the Epicurean term 'clinamen' – the swerve of atoms that brings things in and out of existence – to capture the original position of knowledge. It leaves open the nature and degree of intentionality associated with any particular textual swerve. Specifically, any purported work of 'genius' may be judged the product of either unchecked incompetence or outright plagiarism. Thus, the scholar is a detective who traces the swerves from a present text back to some past text which in turn licenses a judgement about the present text. During the Enlightenment, this function acquired a public face as 'criticism'.

According to the original position of knowledge, the potential for 'information overload' is intrinsically connected to features of the knowledge transmission process itself. For example, in the Middle Ages, it was manifested in the occasions for deviation afforded by the manual transfer of text from one copy to the next. Acknowledging this point does not require taking into account the number of people trying to transmit simultaneously, let alone the motives of these people. To be sure, the presence of more transmitters exacerbates the extent of the problem, which is how we experience 'information overload' today. In any case, the original position is not so different from the variation generated in biological reproduction. In both cases, interesting onto-logical problems arise with the emergence of textual or biological 'mutations' which lie outside a presumed range of expectations. From this standpoint, the Renaissance may be likened to the introduction of a more explicit filter-ing process for textual transmission which standardizes the range of tolerable deviations, very much in the spirit of 'natural selection', understood as the overarching principle of Darwinian biology. Here I mean to include the pro-cesses of editing and curation that serve to determine which texts go forward in time and space.

This policy provides the context to understand Daniel Dennett's (1995) inflated claims for natural selection as a 'universal solvent', not least to the problem of truth. Although epistemologists have protested at Dennett's reduc-tion of truth to a kind of indefinite survival of content, the Renaissance schol-ars saw themselves as doing truth's bidding by setting the standard of fitness required for just such a survival. They came to be greatly aided in this task by Johannes Gutenberg's invention of the moveable-type printing press, which promised to reduce – if not altogether eliminate – the problem of imperfect reproduction, providing (at least in principle) mass access to standardized and (presumably) authorized texts, which could be simply cited without any pre-tence of content reproduction. I shall shortly turn to the ideational implica-tions of this development, but let us first dwell a bit longer on the material implications.

It is easy to see how such an increasingly normative attitude to the brute transmission of knowledge would instil the sort of ethic of defiance that in the nineteenth century came to be associated with Romanticism, the period from which most of the poetic examples invoked in Bloom (1973) are drawn. In short, what an editor might otherwise wish to ignore, suppress or discard as mere error comes to be valorized as *originality*. At the same time, in the modern period *priority* comes to be seen as the complementary 'establishment' virtue, one grounded in a clear sense of a 'source text' in terms of which allegedly 'original' texts may be deemed accretions, amplifications, deviations or errors. The modern idea of *intellectual property* is founded on this dual sensibility, with originality captured by copyright and priority by patent (Fuller 2002: chap. 2). Originality is judged in terms of whether an 'author' is true to oneself, whereas priority is judged in terms of whether an 'inventor' is true to others. To be sure, every Romantic author would like his or her original works to serve as the foundation for others to follow, and likewise every entrepreneurial inventor would like others to presume that by virtue of having come first, the invention sprang out of his or her mind with minimal precedent.

Elizabeth Eisenstein (1979) famously observed that these nascent modern attitudes were given full voice with the printing press, specifically by generating a world of books – and later, journals – that left more time and space for authors to develop and express their own original contributions. Perhaps the first academic work to take full advantage of this technological innovation was Newton's *Principia Mathematica*, a formidable text published in 1687, which presents an account of the 'world system' from Euclidean first principles, with relatively allusive reference to precursors and opponents. In effect, Newton had 'synthesized' (in the Hegelian sense of 'reframed') all that had come before him and reinvented it as a seamless whole, which was presented – at least implicitly – as how the world works from the standpoint of a Creator who uses the minimum number of principles to generate the maximum range of phenomena. To put the matter crudely but not inaccurately, Newton provided a vision of physical reality from the standpoint of the 'Mind of God', the destination which Saint Bonaventure, an early minister general of the Franciscan order and Parisian sparring partner of Thomas Aquinas, had explicitly proposed for all Christian learning four centuries earlier.

Two rather divergent lessons followed from Newton's exemplary achievement, both of which are instrumental for understanding contemporary concerns relating to information overload. Roughly speaking, one relates to the humanities and the other the sciences.

The humanistic lesson relates to the principled availability of all previous scholarship, as epitomized in the proliferation of libraries as searchable repositories of knowledge. This increased the burden of proof on any prospective author to provide 'added value' to any new text produced – a burden beyond simply proving 'originality'. This became a spur to 'creativity' as a mode of self-expression. It acquired enormous political import in Newton's youth through the offices of Oliver Cromwell's foreign secretary, the great poet John Milton, who argued in *Areopagitica* that one should be willing to die for self-expression as the ultimate right of any free person. In effect, the right to free expression was the right to a spiritual life in a world quickly becoming crowded with other such expressions, as both writing and publishing were made more widely available – and where more direct measures were being taken to cull the surplus (e.g. the burning of books and sometimes their authors). But fast-forward 300 years: the advent of the 'postmodern condition' has turned 'self-expression' into a relatively bland, even condescending, phrase for a naive understanding of the multiple and typically conflicting group identities – also known as 'intersectionality' – that contribute to the construction of any individual 'self'. Whereas Milton was mainly concerned with upholding the expressive capacities of the self, the postmodernist is now concerned with whether there is a self that could bear such privileged capacities.

In Milton's day, the point was to liberate oneself from the collected voices of the past which had been suddenly resurrected in all their polyphonic force with the Renaissance and the printing press. In authoritarian hands – be they political or more strictly academic – this textual avalanche could inhibit the capacity of people to manifest their godlike creative powers, in line with the Biblical definition of humans as created *in imago dei*. These creative powers should be understood in the spirit of 'self-legislation', to recall the phrase that Kant would use a hundred years later for our moral sense, based on the idea that we bring order to our unique inner worlds just as God brings order to the outer world in which we are all equally located. This sense of creativity is still honoured whenever we speak of the 'worlds' conjured up by authors and other artists. Nelson Goodman (1978) was only one of several more recent philosophers to catch a glimpse of this conceptual intimacy of 'art' and 'world', which perhaps had received its most full-throttled expression in Friedrich von Schiller's *Aesthetic Education of Man*, written in the wake of Kant's *Critique of Judgement*.

The relevant move that Schiller adapted from Kant pertains to the nature of 'matter', understood as the stuff out of which creation happens. Seen from a temporal standpoint – that is, our own but not God's – matter exists prior to the act of creation as both a means and a source of resistance to our ends. The key property of matter here is its indifference to our ends, as in the 'burden of the past', whose biological correlate might be 'genetic load'. Here Kant followed the ancient Greek practice of treating matter as 'indeterminate' prior to the imposition of form. 'Indeterminate' is meant to suggest 'raw material', as marble, say, is to sculpture. In this respect, the historical record may be treated as a certain disposition of textual matter which is indifferent to whatever ends we might orient the production of our own texts. In other words, the past is less something that we build upon than try to rework to our own advantage. In any case, the moral is clear: to remain open to the future, we are forced to appropriate the past, but the exact nature of that appropriation is left to our own discretion, very much in the manner of a judge deciding a case in terms constrained by statutory and common law. Nevertheless, the result sets a 'precedent' which contributes to the indeterminate textual burden imposed on our successors.

Perhaps the most artful demonstration of this mode of humanistic self-expression – and very likely a source for Kant and Schiller – was to be found in Rousseau's textual practice, especially his bestseller *La Nouvelle Heloïse* (Darnton 1984: chap. 6). Rousseau did no less than extend our sense of 'humanity' by providing linguistic expression to states of consciousness – or 'private life', in the broadest sense – only glimpses of which had been subject to formal representation (e.g. in the plays of Euripides or William Shakespeare), and then often in an especially irrational light. Indeed, William James's early twentieth-century characterization of consciousness as a 'blooming, buzzing confusion' may be understood as an updated version of the 'mind qua raw material' position from which Rousseau started. In turning raw mentality into a well-marked 'mindscape', Rousseau took seriously that language is a species-defining feature of Homo sapiens, a mark of our having been endowed with the divine *Logos* as not only creative but also self-creative.

To be sure, Rousseau stressed, à la original sin, we have historically used language to enslave ourselves by accentuating differences amongst individuals who are in turn arranged hierarchically for purposes of governance. It has meant that people are identified in terms of their assigned roles, which in turn implies that if they wish to say something that breaks with those roles, they must assume the guise of a role suitable for expressing it. Theatre had proved effective in Rousseau's day in voicing disruptive, even radical sentiments (often expressed as irony or satire) in a public space which was specifically protected as 'entertainment', a term whose meaning straddles deliberation and play (Sennett 1977). Voltaire and Gotthold Ephraim Lessing were the living masters of this Enlightenment art. In the process, they helped craft a sense of good prose style in the French and German languages. However, Rousseau was having none of it. Instead, he forged a mode of language use – one which would become standard in novel writing in the nineteenth and twentieth centuries – in which characters expressed their innermost thoughts and feelings, often in quite digressive form, which added an air of psychological authenticity to the presentation.

A key feature of Rousseau's sense of authorship was that he never declared whether he was writing 'fact' or 'fiction' but rather simply took responsibility for the veracity of the thoughts and feelings being expressed and invited the reader to try the content out for themselves, resting his integrity as an author on the consequences of that encounter. In this rather clever way, he evaded those who might be otherwise preoccupied with the source-level accuracy of, say, *La Nouvelle Heloïse* (which is composed as a long and intimate correspondence), without having to pre-emptively neutralize the reader's response by openly declaring his work a purely imaginative construction (which was standard novelistic practice). In light of Rousseau's seemingly lifelong attempt to refashion his Genevan Calvinist upbringing into a radical secular ethic – one which bore most fruit when taken up by the early Marx – the obvious precedent for his textual practice is the personal reading of the Bible on which the Protestant Reformers insisted, which propelled the first spurt of literacy in the modern era. Unlike the Thomists and other 'Doctors of the Church', the Reformers presumed that the Bible was addressed directly to everyone without need of extensive scholarly mediation. Indeed, much of that scholarship appeared to dilute or otherwise equivocate on what the Reformers took to be the Bible's core message.

Given today's politics of knowledge, at least in secular Western societies, it may be difficult to appreciate the role of 'literalism' in the original drive for mass literacy, something which had not previously been a priority in Christendom despite its grounding in a sacred book. The Reformers were clear that even readers lacking sophisticated scholarship could use the Bible to inform their lives. This was the spirit in which they embarked on translations of the sacred book, typically replete with visual images, into the emerging 'vulgar' tongues that seeded the modern European languages (Pettegree 2005: chaps. 5–6). Indeed, modern translation theory – of the sort which influenced Kuhn's (1970) notorious thesis about the 'incommensurability' of scientific paradigms – takes its point of departure from considering how an often extensively rewritten Bible in the newer language (e.g. the King James Version in English) might be understood as somehow bringing readers 'closer' to the original meaning of the sacred book than had they been forced to learn, say, Hebrew, Greek or Aramaic (Nida 1964; cf. Fuller 1988: chap. 5).

Rousseau was one of the first secular authors to cast his works in comparable terms, ones which invited readers to realize the truths of which he writes by inhabiting his words for themselves, without scrutinizing the empirical basis for what is presented on the page. Just as faith in God grew with one's personal encounter with the sacred book, so too Rousseau attracted a dedicated fan base (including fan mail), as he provided a detailed language for articulating quite specific, typically emotion-laden, mental states which heretofore had not been in general currency. Seen from today, Rousseau was conducting what Gustave Flaubert would later call *l'éducation sentimentale*. But the explosive effect that Rousseau's works had in his own time is better explained in terms of the 'dramatization' of private life: Instead of oneself adopting a mask as an actor, à la 'Voltaire' (itself a pseudonym), in Rousseau the mask becomes oneself, a situation which would normally be associated with a holder of religious office. Thus, Rousseau's influence can be traced throughout the modern era in the increasingly exteriorized representations of normally private mental states, ranging from novels to today's self-help books and soap operas.

The scientific lesson of Newton's achievement shifts the focus from originality to priority of authorship. It is epitomized by the medieval saying which Newton used to characterize his own work: 'If I have seen as far as I have, it is because I have stood on the shoulders of giants.' In its original setting, the quote was meant ironically and perhaps even self-deprecatingly, since it only takes a dwarf standing on the shoulders of giants to see beyond them (Merton 1965). It turns out that I only discovered the lineage of this saying after having written the following, with regard to Kuhn: 'But it only takes a dwarf standing on the shoulders of giants to see beyond them, and from a distance, the dwarf may look like the head of the tallest giant' (Fuller 2000b: xii). The second half of this sentence is worth further consideration. In any case, it was often cited by reviewers as an uncharitable spin on the saying's meaning, even though it would seem to be exactly what the medievals had intended.

Nevertheless, Newton's slogan remains apt for the modern scientific enterprise, insofar as the surest route to publication is to propose an incremental improvement to a well-defined body of knowledge. Thus, the superabundance of scholarship which appears to the naive eye as 'information overload' in fact signifies a collective quest in which predecessors focus the prospects of successors.

In this respect, academic publishers do not get the credit they deserve for the promotion of knowledge as a public good (Fuller 2016a: chap. 2). After all, their common online practice of allowing free access to a journal article's abstract, keywords and reference list – charging only for access to the main body of the article – is often sufficient to grasp the contribution that the article makes to the state of knowledge in a given field. In terms of Newton's motto, publication simply marks the success of the first of many rival dwarves scrambling to the top: if dwarf A had not got there first, dwarf B would have got there soon enough. Indeed, this tendency could be rendered demand driven as an epistemic version of the board game Scrabble, which rewards the first journal article that competently assembles a specific combination of references, where 'competence' is understood broadly to include both new findings and clever reasoning.

Perhaps such a contest could qualify for Hermann Hesse's *The Glass Bead Game*. Far from frivolous, this suggestion starts to acquire a larger significance if one considers the vast bulk of scholarship which constitutes what Don Swanson (1986) dubbed 'undiscovered public knowledge', that is, research already published which remains underutilized, if read at all. The idea behind the phrase, which comes from the field of library and information science, is that incentives could be provided – and search engines developed – for scholars to probe the already existing academic knowledge base more deeply to find hidden, possibly cross-disciplinary connections which might then help economize on resources that would otherwise be used to conduct a line of enquiry from scratch. Ideally, people encouraged to try their luck with the literature in this fashion may be inhibited from reinventing the epistemic wheel, a routine feature of research grant proposals, in which researchers review the literature less as a proper resource than as a rhetorical device designed to clear a conceptual space for their own prospective contribution.

However, once we turn the concept of information overload in the direction of undiscovered public knowledge, it starts to look less like an unintended consequence of success than a symptom of outright failure. In this context, a pioneer of digital political campaigns, Clay Johnson (2012) has coined the phrase 'information obesity' to describe the tendency of digital consumers to manage the plethora of information not by regularly sampling across the full range available but by bingeing on sources that reinforce their opinions, a tendency which is facilitated as search engines become increasingly sensitive to digital consumption patterns. As a result, digital consumers may feel very well informed, yet they may be actually ignoring most of the available information. This would be like the full feeling after a calorie-laden fast food meal, from which vital nutrients are nevertheless missing. The lawyer Cass Sunstein (2001) has famously discussed this matter in terms of 'echo chamber effects' which potentially undermine any prospects for digital democracy. Applied specifically to science as a collective enterprise, this phenomenon is recognizable as the 'path dependency' associated with what Kuhn (1970) called 'normal science', namely, the increasingly specialized modes of enquiry which make it ever easier for dwarves to stand on the shoulders of not only giants but also other dwarfs.

At a deeper level, the metaphor of 'information obesity' is very apt. It suggests that the mass of uncited and unread scholarly publications constitutes an idle stockpiling of concentrated intellectual energy. In that case, one may characterize the task ahead in one of two ways, each tied to a different strategy for diagnosing and solving depressions in classical political economy: 'Ricardian' and 'Malthusian' (Fuller 2002: chap. 2).

The Ricardian sees information obesity as symptomatic of a lack of new knowledge to displace or otherwise 'creatively destroy' existing knowledge by rendering it obsolete. This is analogous to curing obesity through increased physical exercise. In Kuhn (1970), 'scientific revolutions' periodically perform this function, in which knowledge cast in the new paradigm effectively metabolizes what is of value in the old paradigm while discarding much, if not all, of the old paradigm's original theoretical framing. In contrast, the Malthusian deals with information obesity more in the spirit of a dietician who recommends eating less and more carefully. Here we find a somewhat different sense of 'creatively destroying' knowledge, namely, by returning to first principles, what Martin Heidegger called *Lichtung* (clearing), which also figures in pedagogy designed to introduce students to a complex body of knowledge without having to reproduce its entire intellectual paper trail. In this way, the appearance of information overload is demystified, if not deconstructed, as each new generation of enquirers is invited to reopen seemingly settled questions via a simplified route to the current state of play.

Up to this point we have been considering information overload at the systemic level. However, at a more micro level, the issue looks somewhat different. In the social science literature which sees 'information overload' as about managing the cognitive resources of individuals, the problem is posed in terms of the information one needs to 'optimize' decision-making. 'Optimization' assumes that there are countervailing forces defining the decision space which are in need of joint maximization. In this case, it means that one cannot delve into every piece of information as much as one might like, but only insofar as it actually facilitates decision-making. For the last 50 years, an inverted U-curve has captured this intuition (Eppler and Mengis 2004). Implied here is that one can become exposed to 'too much information' and hence suffer a kind of fatigue analogous to that of the overworked human body, which may even turn into 'learned helplessness' which renders one incapable of deciding for oneself (cf. Rabinbach 1990; Fuller 2012: chap. 5).

Genealogies of the concept of information overload, while normally begin-ning with Alvin Toffler (1970), eventually hit upon Arnold Gehlen's (1988) concept of *Reizüberflutung*. This Nazi-era coinage is usually translated as 'sen-sory overload', which in the consumer boom following the end of the Second World War came to be associated with the numbing effects of mass advertis-ing, propagated increasingly over the 'cool' medium of television (McLuhan 1951). However, unlike the inverted U-curve purveyors in today's management literature, this lineage is less oriented towards finding just the right amount of information needed for decision-making than towards a general downgrad-ing, if not rejection, of information as a necessary component of decision-making. The various turns against technocracy from the 1920s to the 1960s, often depicted in crypto-Nietzschean terms as 'back to authenticity', 'human emancipation' or 'self-reliance', capture this sensibility. The 'will' often figures prominently in these accounts as doing the work that the technocrats would require of the 'intellect'.

However, one needs to be careful in how this revolt is cast. It would be truer to say that the target of the revolt is not so much information altogether as *existing* information. In practice, the revolt takes the form of a risky interven-tion, itself designed to produce *new* information, which in turn may or may not improve one's capacity to make further decisions, as it reveals something luminous about ourselves or the world. A fail-safe version of this policy is what the political scientist-turned-cognitive scientist Herbert Simon (1947) called 'satisficing', namely, a decision – even a suboptimal one – that is 'good enough' to get you to the next decision.

Seen in these terms, the social engineering experiments promoted by the UK Fabians and the US Progressives, which reached their peak in the hey-day of the welfare state, count as interventions in this 'satisficing' sense, even though the people behind them turned out to be of just those technocrats who were demonized by radicals in the 1960s. Nevertheless, both the technocrats and the radicals shared a scepticism – if not antipathy – to the proposition that the human condition is necessarily advanced by assuming that there is already existing information which once found would provide the solution to whatever problem we face. Indeed, the contrary assumption is at play: the qualitative difference between the past and the future – a normative as well as an empirical point – means that no matter how much information we may have at our disposal, there will always be uncertainties which can only resolved by action, and it is the information which we get from action that is the most valuable of all (cf. Duff 2012: chap. 4).

There is a deeper source for this idea, which may be illuminated by considering the difference between how the humanist and the scientist approach the historical record as an 'information base'. The humanist is likely to dwell on the value orientations implied in the data points of the past. In other words, were we to take each bit of information in its own terms, we might be inclined to orient our own decision space differently. This helps explain the default relativism of much humanistic enquiry, such that even humanists who claim to uphold absolute or perennial values are likely to couch judgements about 'lessons of the past' for today's problems in somewhat equivocal language, reflecting the many contradictory value lessons that history teaches. In contrast, scientists treat the historical record much more instrumentally, albeit in a rather narrow and crude sense of 'instrument'. I say 'narrow and crude' because the competent use of an instrument normally involves an awareness of its side effects and unintended consequences. Yet scientists rarely know enough about the larger contexts in which their chosen historical precedents are embedded to anticipate such matters. Thus, while much bolder in projecting portents from the past, they are also prone to massive 'false positive' effects as they overvalue the significance of the superficially resonant cases across time that they alight upon.

Despite these differences, common to the humanistic and scientific approaches to the historical record is the recognition that a superabundance of possibilities for meaningful action is presented by the available information. In that case, our freedom is expressed by taking a binding decision that removes most of these possibilities from contention in defining the projected future. Now, this 'decision' may be conceptualized in rather different ways, all of which were given ample voice in the debates over science's *Werturteilsfreiheit* ('freedom from value judgements') in the first half of the twentieth century (Proctor 1991: part 2). I shall survey these briefly before drawing some general conclusions about what I called at the outset the 'original position of knowledge'.

Let us start with the sort of conventionalism championed by, say, the Hilbert programme in mathematics which would start from virtually any consistent set of 'axioms' (a term rooted in the Greek for 'value' in the sense of 'worth') but would then judge their informativeness by the psychological novelty of the conclusions which follow by logical deduction. After Hilbert, the logical positivists adopted a more generalized understanding of such axioms as theoretical assumptions from which hypotheses could be derived and then tested against observations, ideally through specially designed experiments – a point which the renegade positivist Popper in particular stressed.

Max Weber, with whom *Werturteilsfreiheit* is most closely associated, was perhaps most explicit in suggesting that the moment of decision could be delegated an experimental outcome or some other 'decisive' observation, which would be 'value neutral' in the sense of free of bias from any of the available value-laden possibilities directly under consideration. In contrast, Weber's younger contemporary and later Nazi-leaning jurist Carl Schmitt was adamant that delegating decision-making to some 'fair' procedure (and here he goes beyond experiments to elections) makes no sense without an automatic means of enforcing the decision. Thus, whereas Weber (as well as the positivists and Popperians) had been very careful to distinguish the prerogatives of what he called, in a set of memorable lectures, 'science' and 'politics', Schmitt put them back together in a way that indirectly provided inspiration for twentieth-century totalitarian politics. Thus, with Schmitt we get the purest statement of information as both distillate and precipitant of action.

The preceding set of reflections on the concept of 'information overload' reminds us that when people speak of the 'pursuit of knowledge for its own sake', the emphasis should be placed on *pursuit* rather than *knowledge*: knowledge itself is not something worth having for its own sake. What I have called the 'original position of knowledge' is one in which we are always on the verge of information overload, simply as a consequence of the processes by which knowledge is reproduced across time and space. These processes are not naturally self-limiting or self-selective. Other mechanisms need to be introduced to take relevant 'decisions', be they at the level of a scholarly editor engaged in quality control or that of the harried administrator operating on a tight schedule. These decisions determine what is good/bad, true/false, useful/useless. Once this general point is admitted, then the significance of two central motifs in the modern history of knowledge production – originality and priority – come into focus, the one characteristic of the humanistic and the other the scientific response to the problem of information overload. In neither case is the problem 'solved'; rather, it is simply organized and managed, as the processes of knowledge production continue to flow.

Chapter 5

SCIENCE CUSTOMIZATION: A PROJECT FOR THE POST-TRUTH CONDITION

Protscience: Science Upfront and Personal

In a post-truth world, the very idea of a 'scientific consensus' may be regarded with suspicion, but science as such is by no means dismissed. On the contrary, it is taken personally. This change in attitude compares to the shift that occurred during the Protestant Reformation, the moment when Christianity ceased being a unified doctrine delivered with enormous Latinate mystique from a 'high church' pulpit. Thereafter it became a plurality of faiths, whose followers stake their lives on their own distinctive understandings of the Scriptures. In the case of science, I have dubbed this process *Protscience*, short for 'Protestant Science' (Fuller 2010: chap. 4), by which I mean to include a pattern evident in the parallel ascendancies of, say, intelligent design theory, alternative medicine and *Wikipedia*.

The Protestant Reformation was the first step on the road to the secularization of Europe, which Max Weber famously described as the 'disenchantment' of the Western mind. Protscience's relationship to this process is ambiguous, as it both disenchants scientific authority and re-enchants science itself as a life-shaping form of knowledge (Fuller 2006b: chap. 5). In any case, Protscience takes very seriously the idea that any form of knowledge that aspires to universal scope for its claims must have universal appeal for its believers. In more fashionable language, it involves the reflexive shaping of self and world to enable one to live – or die, as the case may be – with whatever one happens to believe.

Interestingly, just as the original Protestants were demonized by Catholics as 'atheists' for their refusal to defer to papal authority, today's Protscientists are denounced as 'anti-science'. In both cases, however, the people concerned are generally well educated and quite respectful of the need to provide reasons and evidence for their beliefs. Not surprisingly, then, Protscientists make much of the hypocrisy of established authorities that fail to live up their own avowed epistemic standards. Thus, any report of scientific fraud is grist for the

Protscience mill, yet one more reason to take matters of knowledge into one's own hands before the entire enterprise of enquiry becomes corrupt.

Unsurprisingly, just as Catholic theologians said of Protestant readings of the Bible during the Reformation, scientists have complained that this availability of information has only served to foster misunderstanding and charlatanry. Without wishing to deny that possibility, it is also true that the public appropriation of scientific facts and concepts, however wrong-headed or bizarre it may seem to professionals, results in a public more willing to take personal responsibility for the decisions they make about whether to carry an umbrella, invest in a company, undergo a treatment or, indeed, evacuate a town. This permits scientists to speak freely about their research without fear that they might be held liable for the consequences of what they say. In effect, the interpretive burden has been shifted to a presumptively engaged and intelligent audience.

This brave new world of Protscience is the latest phase of secularization, whereby science itself is now the target rather than the agent of secularization (Fuller 1997: chap. 4; Fuller 2000a: chap. 6). Nowadays the Protestant Reformation of sixteenth- and seventeenth-century Europe is taught as an important episode in the history of Christianity, but it also marked the first concerted effort to democratize knowledge production in the West, specifically by devolving religious authority from the Church of Rome. Indeed, the formal separation of knowledge production from the reproduction of social order is perhaps secularization's strongest institutional legacy, which began with the political separation of church and state. We are now entering the second period, consisting in the devolution of the dominant epistemic authority of our time – science – from the state-based institutional privilege that it has enjoyed, say, since the founding of the Royal Society of London.

Readers who remain sceptical of this development are invited to consider some signs of this sea change in the public's engagement with science. Most obviously, science's increasing visibility in public affairs has coincided with the ability of people to access the entire storehouse of scientific knowledge from virtually any starting point on the Internet. The result has led to a proliferation of what used to be called (sometimes derisively) 'New Age' science hybrids, some of which have breathed new life into movements previously thought defunct, including creationism and homoeopathy. Here it is worth recalling three general senses in which science is inherently democratic. However, it is not clear that all of these senses are mutually compatible:

1. Science is 'universal knowledge' in the strict sense, which it to say, it covers all things for all people. This is the spirit in which the so-called reductionist impulse in modern philosophy – most clearly associated with both

nineteenth- and twentieth-century positivism – should be understood. The appeal to reason and observation (or logic and experiment) in Auguste Comte was designed to demystify theology's epistemic authority. In the Vienna Circle, it was designed to demystify lay academic expertise. Moreover, the latter positivism was dedicated to correcting the excesses of the former positivism, since Comte would have science become the new church – instead of preventing the rise of new churches. Popper's 'science as the open society' was a radical expression of this sentiment.

2. Even though until relatively recent times the university has been populated by elites, the modus operandi of the university has been 'democratic', given the Humboldtian mandate to integrate original research with mass teaching. Thus whatever initial advantage is gained from an innovative piece of research is minimized if not eliminated once it is made available for more general consumption in the classroom. In this respect, teaching may be seen as *epistemic entrepreneurship* in the Schumpeterian sense of being engaged in the 'creative destruction' of expertise by enabling students to spend less effort than their original researchers did when they first acquired the same knowledge (Fuller 2009: chap. 1; Fuller 2016a: chap. 1). The textbook is the great symbol of this process, and not surprisingly its introduction across the sciences in late nineteenth-century Germany corresponded with rapid nation-building ambitions.

3. There is also the inherently 'democratic' nature of the scientific role or personality. This can be understood in three separate senses: (a) universal subordination to a common disciplinary ideal or paradigm (i.e. all scientists are equal in their search for truth, implying similar capacities for insight and error); (b) each person's equally individual embodiment of universal knowledge (i.e. the romantic German ideal of *Bildung* championed by Johann Wolfgang von Goethe); (c) collective production and ownership of knowledge as a public good, which implies that there is a scientific standpoint, in principle accessible to everyone, that can be used to govern society (i.e. the ideal common to positivism, utilitarianism and dialectical materialism). These three senses of 'democratic' correspond to the nineteenth-century self-understanding of science in, respectively, Britain, Germany and France (Fuller 2010: chap. 3).

The printing press was crucial to the first rise of Protestantism as both a convenient and a lucrative medium for people to acquire the cognitive resources needed to decide for themselves what to believe, and thereby no longer defer to the local priest. Indeed, with the Bible leading the way, books became big business once they were published in the vernacular and in a portable form that allowed for easy circulation. This tendency accelerated during the

Enlightenment, and is normally credited for the comprehensive liberalization of Western culture (Wuthnow 1989). Over the past quarter-century, a new wave of vernacular publishing of equal significance has been made possible by computer-based information technologies – from web searches to social networking sites. Its impact on the distribution of epistemic authority in society is palpable, though its long-term consequences are unclear. However, it is certain that the institutional solutions for managing the diversity of opinions and claims to legitimacy that were developed in the wake of Protestantism – the secular state and the scientific method – are themselves undergoing a 'crisis of legitimation'.

These old solutions were originally designed to resolve potentially violent disputes among the different and often competing interpretations, applications and extensions of the biblical message. In this respect, regular elections and controlled experiments have functioned similarly. Moreover, as the state came not merely to protect but also to promote the welfare of its citizens in the nineteenth and twentieth centuries, science became increasingly implicated in – and defined by – the state's workings. Thus, scientific elites are effectively high priests of the secular state, the ultimate validators of policy-relevant admonitions that range over personal health, the state of the economy and the prospects for the environment. Protscience challenges such close state-science ties, not least because the scientists are unaccountable to those who they govern by proxy.

Protscience aims to re-jig the balance of epistemic power, so that, say, a doctor comes to regard a patient in her clinic as more like a client who needs to be sold on a treatment than a machine that needs to be fixed. This is precisely because Protscientists are convinced of science's integral role in their own lives. For that very reason, they insist on taking an active role in determining how that integration occurs. Thus, they take soundings from alternative, including Internet-based sources and supplement the methodological uncertainties of all scientific research with their own experience and background beliefs. But perhaps most importantly, Protscientists uphold their right to decide scientific matters for themselves because they are the ones who principally bear the consequences of those decisions. This results in a pick-and-mix approach to science that retains the vast majority of accepted scientific fact and theory while giving them a curious spin in light of distinctive explanatory principles and life practices.

The Public Intellectual as the Proto-Protscientist

Perhaps the most obvious precedent for Protscience is the rise of a 'free floating intelligentsia', to recall Karl Mannheim's resonant phrase for the class of critics and journalists who starting in the nineteenth century began to earn a living in a nebulous intellectual space located somewhere between the academic and political spheres (Fuller 2005b). To this day, academic experts all too readily condemn intellectuals simply because they manage to profit from expert knowledge either without having acquired the proper credentials themselves or – perhaps more damningly in expert eyes – without requiring others to do so. In other words, the intelligentsia are by nature 'epistemic trustbusters' who by routinely operating in multiple media are forced to develop keen powers of discrimination that enable them to spot and circumvent the rent-seeking tendencies in academic thought as they craft a message capable of striking their audience in just the right way.

Intellectuals normally begin life among academics. But they quickly realize that the specific means by which academics produce and validate knowledge speaks more to how the academics themselves came to know something than what others should make of that knowledge in the future. Thus, an intellectual pondering a sound bite or pitching a newspaper column is thinking about how to appropriate academic knowledge most effectively, perhaps even in the spirit of Pierre-Joseph Proudhon's 'Property is theft', in which academics appear as epistemic barons, whose sense of entitlement outweighs their actual custodial role in the maintenance and transmission of knowledge. While intellectuals have generally not tried to demystify or deskill academic knowledge, that can be the unintended consequence. When intellectuals hit the mark with a 750-word article (as opposed to a 75,000-word book), they demonstrate that it is possible to elude the rents that academics normally impose as 'educational requirements' to acquire an empowering form of knowledge. At such moments, intellectuals are truly fulfilling their classic 'enlightenment' function in society.

In this respect, intellectuals are entrepreneurs of epistemic efficiency who see academic expertise as channelling the otherwise free-ranging human mind in ways that in the first instance benefit academics, whatever other secondary benefits might accrue to the rest of humanity, say, from students who go off to do other things in life.

A specific form of intellectual life is associated with science journalism, which deserves mention in the context of Protscience, given the tendency – especially among academics – to think that before science and technology studies began to demystify taken-for-granted views about the workings of science, science journalists were nothing more than scientists' press agents. However, the history is more interesting, profound and yet to be written. But the basic outlines are clear: start with H. G. Wells, now known as a 'science fiction' pioneer. This ex-student of Thomas Henry Huxley, Darwin's greatest contemporary defender, was really interested in using recent discoveries and nascent tendencies in science as a vehicle for reporting about how the world might look in the future, a kind of 'anticipatory journalism', which might serve either to hasten or retard the future so projected, once people found out about it. Interestingly, Wells thought that here he was following in the footsteps of the great nineteenth-century progressivist thinkers Comte, Marx and Herbert Spencer, which in turn inspired his unsuccessful bid to become the first UK professor of 'sociology', at the London School of Economics in 1907. Wells called sociology the 'science of utopias' (Lepenies 1988: chap. 5). I remain sympathetic to Wells's original conception of sociology (Fuller 2015: chap. 6). We shall revisit this sensibility in the final chapter of this book, on 'forecasting'.

Of course, Wells was not alone in regarding science journalism as the ideal vehicle to publicize the advancement of science, usually to embrace the science but sometimes to immunize the society. Indeed, until the Cold War, great science journalism was about projecting bold technoscientific futures, with population genetics pioneer Jack Haldane and *New York Times* columnist Waldemar Kaempffert leading the way, the former looking fondly to the Soviet Union and the latter to Fascist Italy. Both eventually rowed back, and the Cold War began a more critical period of science journalism, in which figures such as ecologists Barry Commoner and Rachel Carson stressed, in rather different ways, science's irresponsible complicity in undermining the lives of the people – and other creatures – that it had supposedly aimed to promote. Here were planted the seeds of the still current idea that the scientific establishment is just as corruptible as any other political agent, the 'scientific-military-industrial complex' in the Cold War case. Over the past 40 years, science popularization has swung between those who believe that science can define our fate, for better or worse (contrast the 1970s works of Konrad Lorenz and Richard Dawkins, both ethologists), and those who believe that science needs to be opened to more perspectives and greater scrutiny. The latter is associated with science and technology studies, but *Scientific American*'s John Horgan (1996) deserves credit for having pursued a similar path within normal journalistic conventions.

With the advent of the Internet, science journalism has become more like investigative journalism, in which the journalist ends up participating in the construction, if not reconstruction, of scientific knowledge. This has given scientific journalism a more unpredictable and even subversive character. Two tendencies have contributed to this phenomenon. The first tendency is a classic 'supply push' and 'demand pull' dynamic: in this case, a surplus of scientifically trained people spilled over into journalism just when the public has come to think of itself less as spectators than consumers of science. Thus, non-scientists wish to know from the science journalist whether the scientists' products are worth buying. Take Ben Goldacre. Despite being an Oxford-trained medical doctor and self-avowed scourge of 'Bad Science' (the name of his weekly column in the *Guardian*), his modus operandi involves subjecting scientific papers – mostly in biomedicine where the financial and public safety stakes are highest – to statistical and other research design tests, which end up uncovering flaws even in papers that have passed the peer-review process. Goldacre's campaign has taken him to the doorstep of 'Bad Pharma' (Goldacre 2012). One recalls here Ralph Nader's 'test-driving' cars rolling off the Detroit assembly lines in the 1960s to see if they lived up to manufacturers' claims, which sparked the original consumer movement.

The second tendency contributing to this brave new world of investigative science journalism is exemplified by Evgeny Morozov, perhaps the world's leading critic of Silicon Valley hype, according to which every problem might be solved by improved information technology, a doctrine that he has dubbed 'solutionism' (2013). A Belarus native now ensconced in the Valley, Morozov was a young beneficiary of George Soros's Open Society Foundation, which aims to spread liberal democracy in former communist regimes. But Morozov stands out for having updated the proverbial armchair critic, a well-read and fluent humanist who cannot programme a computer but whose endless scouring of cyberspace enables him to show how Silicon Valley dreams are not borne out by reality. He does this typically by citing (or spinning) text against text. In effect, Morozov is the 'evil twin' of a dedicated Apple user – that is, someone who takes the hype literally only to be endlessly disappointed. Morozov's large following vindicates one of his basic points, namely, that despite the hype surrounding 'open access' in the world of information technology, most Internet users are as techno-illiterate as Morozov and who therefore quite understandably want Silicon Valley computer programmers to be held externally accountable for their claims.

From a post-truth perspective, while scientists might see only epistemic clouds, there is a silver lining. Instead of focusing on the distrust of scientists, the public has become more actively interested in what scientists are doing, even if the public's conclusions go against established scientific judgement. This suggests that the likes of Goldacre and Morozov, from their rather different epistemic starting positions, are pointing to what the UK science communication researcher Alice Bell (2010) has called 'upstream' science journalism, which would report on ongoing research before it reaches the publication stage. Taken to extreme, it could amount to a 24/7 'rolling news' approach to the public's access to scientific research, in which the laboratory becomes the scene of a 'reality television' show. While this prospect might strike some scientists as a nuisance, nevertheless it provides an opportunity for the public to develop personal stakes in the research outcomes.

All of this may sound like opening the door to science being cherry-picked to suit particular world views. And to a certain extent, it is just that. However, a mature secular democracy is capable of respecting even those who wish to embody in their lives hypotheses that scientists have dismissed. I have no doubt that in such a tolerant environment people would continue to fund and consult scientific research. But the conclusions that each person draws from it would be his or her own, for better or worse. Taking science personally ultimately means turning oneself into a living laboratory. The question for a society supporting this activity, then, is the extent to which its members can fairly benefit from these invariably risky individual initiatives. This requires, first of all, decriminalizing the initiatives themselves, and furthermore providing the appropriate compensation and recognition for those whose personal risks may serve to benefit all in what is a fairly described, repurposing William James, as the 'moral equivalent of war' against our collective ignorance (Fuller and Lipinska 2014: chap. 4).

The Science Customer Who Need Not Be a Science Consumer

The marketing literature draws a usefully sharp distinction between *customer* and *consumer*. They are successive stages in the supply chain, where 'custom' refers specifically to the exchange between a manufacturer and a retailer. The customer is, strictly speaking, the client, someone who purchases a good or service, regardless of what she or he does with it. In contrast, a consumer is a customer who actually uses the good or service. While customers and consumers are very often one and the same, it is possible to be one without being the other.

A 'science customer' may purchase some epistemic goods and services without necessarily consuming them. For example, she may learn all about the neo-Darwinian account of evolution and even pass along its content to others without ever believing the account herself. Indeed, I have argued that this is the optimal strategy for latter-day creationists – also known as intelligent design theorists – to make headway in a scientific marketplace in which evolutionary theory provides the only tradable currency (Fuller 2008: chap. 1). To adopt such an attitude is to behave like the retailer who purchases a good to sell someone else without ever consuming the good herself. Conversely, a 'science consumer' may never have intended to ingest the genetically modified organisms that are already contained in most of the foods she eats. Indeed, she may even believe that such organisms are harmful or unnatural. And while her own consumption patterns – especially if she remains healthy – testify against her beliefs, she may nevertheless have legal grounds to sue the relevant food providers for having failed to secure informed consent for her custom.

This customer/consumer distinction is by no means alien to the historical understanding of science. In particular, the image of science as an abstract manufacturing industry that converts raw material (empirical data) into usable knowledge products (laws, solutions, predictions, and so on) without 'configuring' the exact user – to recall the memorable title of Steve Woolgar (1991) – has been always strong. A clear example is what the great social science polymath Herbert Simon in the 1980s began to call 'computational scientific discovery' (Langley et al. 1987), which aims to produce the widest range of known scientific findings from the fewest number of inference rules. This body of knowledge and reasoning would then serve as a platform – or 'fixed capital', as economists would put it – to project an indefinite range of future findings, only a fraction of which could be ever surveyed, let alone adequately pursued by human beings. That Simon called his succession of computer programmes 'Bacon' after Francis Bacon shows that he knew exactly what he was doing in conceptualizing science as in mode of production rather than strict product line terms. Philosopher Paul Humphreys (2004) went further to argue that science might be more efficiently done by such 'computational science machines', thereby removing humans from the manufacturing stage of science and leaving us with the roles of science customers and consumers.

To be sure, this would not necessarily involve humans offloading their brains to machines. Rather, if Simon and Humphreys are right, different cognitive capacities are required of the science customer and consumer than that of the science producer. Humans may excel at those. Such capacities are on display in art connoisseurship, an analogy that Kuhn's mentor, James Bryant Conant, explicitly drew to explain the sense in which non-scientists should have an 'understanding' of the nature of science (Fuller 2000b: chap. 4). Our humanity would hang on that skill set. Moreover, this sensibility is not so very different from what the Nobel Prize–winning chemist Walter Gilbert (1991) projected for bioinformatics a quarter-century ago, in which amidst the array of DNA strings, some specific ones would stand out to the canny 'bioprospector' as investment opportunities for new medical treatments. Indeed, he had been following his own advice since the late 1970s when he and some fellow scientists established the commercial enterprise BioGen, which later became the platform for recruiting the people would become the founders of the Human Genome Project (Church and Regis 2012: chap. 7). It was this trajectory that eventuated in one of their number, Craig Venter, becoming known as a 'venture capitalist' of the genome.

Formalizing the distinction between science customer and science consumer could have avoided the unfortunate situation that befell the six Italian seismologists who (with one politician) were sentenced to six years in prison in October 2012 for manslaughter based on what turned out to be false assurances about an earthquake in 2009 that left 300 people dead, 1,600 injured and 65,000 people homeless in L'Aquila, a district that is normally home to 100,000. To be sure, the scientists stated quite clearly – and accurately, given the best evidence available – that the earthquake was highly improbable. But of course, it is in the nature of improbable events that they happen every so often. Rather more damningly, the experts appeared to have spun this improbability as a counsel of complacency. The judge, whose verdict reflected public opinion, stressed that the severity of the punishment stemmed only from this counsel and not the original probability estimate.

The scientific community was quick to express outrage. The 25 October editorial in *Nature* led the charge, claiming that henceforth scientists would be reluctant to speak their minds freely in public settings, especially ones that might bear on policy. Italy's relatively poor track record in research funding was brandished as a symptom of science's low national esteem, which in turn made the seismologists an easy target for populist rage. However, this analysis itself is a bit too easy, even though ultimately I agree that the fault lies more squarely with the public. But my reasoning differs radically from that of the editors of *Nature*. (It is worth noting that these convictions were overturned by a higher Italian court within two years, by which time media attention had shifted to the Mafia's possible involvement in L'Aquila's reconstruction.)

There is a 'plague on both your houses' character to the unfortunate turn of events at L'Aquila. No doubt at work here was a paternalistic arrogance all too common among scientists that makes them forever susceptible to political manipulation. In this case, the scientists assumed that they knew best how to interpret the data, and so, prodded by politicians, they stressed the lowness of the probability of catastrophe to compensate for what they believed would have been otherwise an irrational public response. But is the public not entitled to draw their own conclusions and, if necessary, learn from their own mistakes? Indeed, arguably the lion's share of the blame for this incident belongs to the public, who had unreasonably expected scientists not simply to inform but also to instruct them. Clearly L'Aquila's residents had not taken the 'Protestant' turn in their engagement with science alluded to above.

Historically, the scientific community has tended to act most 'politically' in the context of protecting the autonomy of their research. Claims to autonomy have extended to the topics they worked on, the methods they used and whatever conclusions they might reach. Securing these claims usually meant a pact of mutual non-interference between scientists and politicians. This arrangement, as we have seen, was enshrined 350 years ago in the Charter of the Royal Society of London. However, in 1911, Germany established the first institution – the Kaiser Wilhelm Gesellschaft – that linked the fates of science, industry and government in projects of mutual benefit. And as we saw in the previous chapter, Fritz Haber was one of the early notable – indeed notorious – beneficiaries. While the second half of the twentieth century witnessed the spread of these so-called 'triple-helix' arrangements, their original incarnation produced disaster. Germany's belligerent stance in the First World War had the full backing of what had become the world's premier scientific community. Perhaps unsurprisingly, in the aftermath of the nation's humiliating defeat, a profound anti-scientific cultural backlash set in, sowing the seeds of much of contemporary fundamentalism, racism and irrationalism (Herf 1984).

Reflecting on this history, some scientists have called for their taking an even stronger role in public affairs, but this time without being hamstrung by self-interested politicians and businesspeople. The roots of this idea are traceable to a Soviet-inspired 'scientific vanguard' that was developed and popularized in the West by the British Marxist physicist John Desmond Bernal (1939). Nowadays it is presented in more democratic, sometimes even populist terms. Consider *The Geek Manifesto*, a widely discussed call to arms, penned by Mark Henderson (2012), head of public relations for the Wellcome Trust, Britain's largest science-based private foundation. Henderson, previously science editor for the *Times* of London, pursues his own version of the new breed of activist science journalism discussed in the previous journalism. In particular, he believes that the collective intelligence of democracy is raised by proportioning authority according to evidence, such that those who know more should be given a larger say in policy. Stated so baldly, the proposal sounds elitist. Yet that great nineteenth-century liberal John Stuart Mill held just such a view. And the more that failures to follow 'proper' scientific advice can be presented as threats to the public interest, the more persuasive *The Geek Manifesto* appears.

However, as the self-deprecating term 'geek' suggests, the manifesto's target audience is science's petite bourgeoisie – that is, the computer jocks who try to escape their day jobs by reading popular science and science fiction, which fuel their web-based interventions in a seemingly endless war against 'pseudo-science', which often includes encouraging the more extreme rhetoric of Richard Dawkins against religious believers. Whatever else one may wish to say about these people, who no doubt find their lives enriched by engaging in such cyberwars, they are not front-line contributors to the research enterprise. This may help to explain why the leading scientific institutions have not signed up to *The Geek Manifesto*. Indeed, this scientific call to arms may ultimately express a wish that is best left unfulfilled.

One aspect of politics that tends to be neglected in discussions of *The Geek Manifesto* is what should happen in the event that Henderson's newly empowered scientists get things horribly wrong, as in the recent L'Aquila earthquake case. To be sure, the verdict came under heavy fire from the world's scientific community. Yet, such outrage suggests that scientists have yet to grasp fully an elementary lesson of democratic politics – that with power comes responsibility. The Italian judiciary originally portrayed the scientists as having abused the trust of the affected residents. And if people are meant to trust blindly scientists speaking from their expertise, then that is a fair reading of the situation. To avoid similar situations in the future, the public should adopt the stance of clients for science, customers who need not be consumers. Such an arrangement may not minimize the likelihood of risky judgements about the world but it would certainly minimize the risk that scientists and the public pose to each other, as each is formally apportioned their own share of responsibility for whatever happens.

The Role of Customized Science in the Future of Democracy and the University

In the age of Protscience, the public would continue to fund scientific research without being bound to the scientists' own interpretation of their findings. They would be science customers without necessarily being science consumers. To be sure, interesting legal questions arise about exactly what scientists should be required to say so that people can draw reasoned conclusions. But in principle these questions are no trickier than those relating to any client-based transaction: the client pays simply to receive relevant information that he or she might not otherwise possess, but is then free to decide what to make of it. Homeowners should be 'free' to ignore the advice of seismologists in exactly the same sense that patients are 'free' to ignore the advice of their physicians – and thrive or suffer accordingly. Once we reach that state of moral parity, then we can claim to live in an enlightened secular democracy in which scientists need not fear that they will be imprisoned for speaking the truth as they see it. That is the utopia envisaged by Protscience.

In this respect, the distinction between the science customer and the science consumer serves to drive a wedge in the still popular, philosophically rationalized inference: *the more science one knows, the more one's beliefs will conform to those of the relevant scientific experts.* In the science communication literature, this inference is often derided as the 'deficit model' for presuming that a simple deficit in knowledge – rather than a difference in the ends for which knowledge is sought – is the main problem with the 'public understanding of science' (Gregory and Miller 2000). Of course, scientists who work in an academic setting where professional advancement depends strongly on peer approval will be susceptible to a variety of incentives and pressures to conform to current expert judgement. Yet, even such institutionalized social control is not guaranteed to work if, say, scientific deviants can find adequate alternative publication outlets. However, the full import of a science customer who does not necessarily consume the science that she or he purchases is best seen in the vast majority of people – including even scientists outside their specialities – who take a much more 'pick and mix' attitude towards the knowledge claims they encounter in science. This includes the following practices:

1. Accepting scientific facts as no more and no less than sociological facts about the collective judgement of the relevant scientists, which is likely to change in the face of new evidence (assuming that the scientists are open to it).
2. Accepting scientific facts as they are, but not granting them the same significance accorded to them by the relevant scientists.
3. Accepting scientific facts and perhaps even granting their ultimate significance, but concluding that they could be explained tolerably – if not equally – well by an alternative theory to that of the scientific orthodoxy.

All three practices can be justified on epistemologically sound grounds, certainly grounds that are sounder than whatever licenses the striking ignorance that people who claim to believe in orthodox science have of the content of that science (Fuller 1997: chap. 4).

A lightning rod issue for science customization is the 'placebo effect' in medicine (Evans 2003). Science customers are well aware of trade-offs involved in relying on clinical trials: their ability to determine the exact physical effects of novel drugs and treatments is offset by complexities in the likely contexts of actual use, where the patient's lifestyle, frame of mind and relationship to the attending physician may enhance, diminish or simply alter the predicted effects. Indeed, drugs and treatments that fail to be robust under variable real world usage have arguably done more harm than, say, homoeopathy and other forms of complementary medicine whose practices involve physically inert substances coupled with psychological uplift from the physician. Unsurprisingly, the sorts of invasive ('allopathic') treatments associated with 'scientific medicine' clearly start to outperform complementary medicine only in the final third of the nineteenth century. At that point, hospital clinics begin to be regularly used as test sites for new treatments, resulting in a systematic record of successes and failures that could enable collective learning to occur in what had been heretofore a largely privatized medical profession (Wootton 2006).

An adequate response to this history requires resisting a knee-jerk philosophical impulse to demonize such science customers as 'relativists' who *merely* appropriate science to bolster beliefs that they would already hold on other grounds. The likely source of this philosophical reflex is the prejudice that 'expert scientists' are concerned with a wider epistemic horizon than 'lay scientists'. In other words, the experts are concerned not merely with what suits their own personal interests but some larger, disinterested conception of truth. Here we need to disaggregate *space* and *time* when we speak of 'wide'. Let us grant space to the experts. In other words, experts very likely issue a measured judgement based on a snapshot of a broader range of perspectives than lay people. But this does not deny that the laity are quite practised in assessing their own long-term prospects, in terms of which scientific judgement can appear quite changeable. Consider someone like myself who was born in the midst of the Cold War. In my lifetime, scientific predictions surrounding global climate change have veered from a deep frozen to an overheated version of the apocalypse, based on a combination of improved data, models and, not least, a geopolitical paradigm shift that has come to downplay the likelihood of a total nuclear war. Why, then, should I not expect a significant, if not comparable, alteration of collective scientific judgement in the rest of my lifetime?

To be sure, such a 'pessimistic meta-induction', as Hilary Putnam (1978) memorably called it, is not guaranteed. However, the historical precedent may serve to motivate people to participate in the scientific enterprise, especially if their interests would stand to gain by a paradigm shift in how the science is done. Thus, creationists who take seriously the idea of a 'young earth' reasonably study the radiometric techniques used to date events in geological and cosmological time, albeit for purposes of showing their flaws. Ideally the efficacy of such study will be borne out by research that impresses peers. Depending on the extent to which scientific authority devolves in the future, publication in other forums might serve equally well to sway the relevant minds.

Whatever else one might wish to say about 'science customers', they assume responsibility for their science-based decisions. They are not ignorant consumers, as demonstrated by their explicit yet circumscribed deviation from the scientific norm. Here it is worth acknowledging the various reasons why one might be a customer but not a consumer. Perhaps the oldest historical reason relates to the social integration of deviant classes or deviant practices. Here the process of abstracting goods from their normal contexts of use that characterizes exchange relations – that is, the conversion of value to price – facilitates the comparison of the previously incommensurable. Thus, when offered a cow in trade, I need not evaluate it purely in terms of my personal use (e.g. do I like beef or milk?) but consider it as something that may be traded for something I really could use. Similarly, a creationist may invest in a science education because she can trade on that to promote her own world view in some way or other (e.g. someone who becomes expert in radiometric geology and cosmology to overturn the status quo), but equally she might acquire a science degree simply to gain credibility in public debate.

In addition, a sharp customer-consumer distinction also enables the individual to acquire epistemic authority by extending the range of choice enjoyed by others instead of imposing a particular world view on them. In this respect, Max Weber's famous defence of free enquiry for both academics and students imputes to the lecturer the role of intellectual retailer who comes to be respected mainly for his range of attractively displayed epistemic offerings that entices students to make judgements about matters that they might not have otherwise thought about. Finally, the customer-consumer distinction creates opportunities for internal trials of faith, the result of which should somehow leave the individual stronger. I say 'somehow' because diverse responses may follow, including these: (a) the customer is converted to a consumer (what in the past might have been considered the default position); (b) the customer is immunized against being a consumer (e.g. a creationist who accepts at least some of the evidence for evolution but manages to contain its effect on her world view, if not give it a creationist spin); (c) the customer may acquire a clearer understanding of her refusal to consume (i.e. the cognitive import of resistance to temptation).

Three specifically political-economic challenges face the democratization of science in an era of Protscience that could pervert the course of enquiry:

1. Science comes to be evaluated in the same cost-accounting terms as other large-scale public and private enterprises. This could lead to a massive shift in what are counted as the costs and benefits of scientific research. For example, science is normally evaluated in terms of very generous time frames that ignore opportunity costs and collateral harms. Thus, a large initial investment that diverted resources from the development of other alternatives is often credited as an overall benefit because enough time is allowed to pass – which enables that initial investment to interact with other investments – so as to result in a worthwhile outcome. Moreover, whatever harms are caused along the way are presumed to have been avoidable and not a necessary feature of the investment strategy. This abstract scheme arguably captures the history of nuclear physics research since the 1930s. But even in cases where a small initial capital outlay eventually produced major benefits – as Newton's work did vis-à-vis the Industrial Revolution or Michael Faraday's vis-à-vis the electrification of world – these outcomes may have been overdetermined, such that even without Newton's or Faraday's particular contributions, the benefits would have happened anyway by some other means. In that case, one might query the extent to which a particular scientific theory or research programme is necessary to make 'progress' in the sociologically broadest sense.

2. The idea of science as a public good has been most persuasive in terms of a Keynesian logic, whereby the benefits of scientific knowledge are seen as 'multiplier effects' from a relatively limited initial investment. Thus, in the welfare state's heyday, all taxpayers subsidized universities that only 10–20 per cent of a generation's cohort would attend. It was presumed that within a generation – if not sooner – those few would produce discoveries, create businesses, open vistas and the like that benefit everyone in society. However, once everyone is expected to attend university in order to get a job credential, taxes rise, patience shrinks and frustration and disappointment inevitably set in. Moreover, the expectation that universities are the de facto gatekeepers of the labour market means that they have not only forfeited their own autonomy to those whose success depends primarily on short-term adjustments to the market (i.e. employers) but also signalled to primary and secondary school educators that any deficiencies that they failed to address adequately will be remedied at the university's tertiary level. The result is 'credentials creep', whereby more academic credentials are required to gain comparable advantage in the labour market (Wolf 2001).

3. To ensure that the democratization of science does not to slip into sheer marketization, knowledge as a public good must be expressly manufactured and not simply presumed to emerge naturally. This means that what people *need* to know must be defined in terms – and driven by a standard – other than what they *want* to know. Otherwise, the value of knowledge will become a short-term survival strategy that promises the most gain at the least cost. It is worth stressing that the problem here is not the simple drive towards 'efficiency' in knowledge production implied in such a strategy, but rather a failure to recognize that efficiency only makes sense relative to particular goals with specific time horizons. To assess efficiency vis-à-vis knowledge manufactured as a public good, one needs to do more than ensure that those paying up front receive a positive return as soon as possible. This point applies to both impatient investors in innovative research and students paying for university degrees. In both cases, people need to be instructed, incentivized, constrained and nudged to think of knowledge as providing less direct benefits doled out over a longer term. Only then can one appreciate the value of knowledge as a public good (Fuller 2016a: chap. 1).

Science's customization was made possible in the same way as science's universalization, namely, by the transfer of scientific authority from a specific body of people who acted as a guild to an abstract method that could be deployed in principle by anyone to any end. Bacon perhaps unwittingly triggered the process by envisaging a state-supported House of Solomon that would produce science for the public good. However, because Bacon was in no position to determine exactly who would constitute this House or how it would be institutionalized, he effectively defined science at a level of abstraction that permitted multiple realizations. What is reasonably clear from Bacon's own writings is that the pursuit of science was partly about rational psychiatry (what René Descartes called 'rules for the direction of the mind') and partly about judicial review (what Carnap called 'criteria of testability'), all in the name of sublimating potentially endless metaphysically inspired disputes in a manner that would be binding for all parties. In this respect, the scientific method would provide a common currency for the transaction of otherwise incommensurable knowledge claims.

What I have just described requires that any method worthy of the name 'science' be neutral with respect to the knowledge claims that it assesses. For Bacon, the substantive ends to which the scientific method would be put would not come from the scientific community but from the politicians. This point is worth stressing, for while the Royal Society is normally presented as based on Baconian principles, its corporate charter made it completely independent of state control, perhaps reflecting its founders' scepticism about a sense of political sovereignty that is at once absolute and experimental in orientation. (Here recall Hobbes's career trajectory, starting as Bacon's private secretary and ending up as persona non grata at the Royal Society: Lynch 2001). Nevertheless, the logical positivists tried to turn Bacon's sense of neutrality to great effect by outlining various universal logics – both deductive and inductive – of empirical assessment. Popper famously saw the matter in more idiographic terms, drawing specifically on Bacon's idea of a 'crucial experiment', whose sense of adjudicative neutrality rests on the construction of the ultimate trial, the outcome of which would clearly divide the fates of two rival hypotheses.

All of these developments in what philosophers call the 'demarcation' of science aided science's customization by allowing people holding different world views to see their relative public epistemic standing at any given moment, with an eye to improving it. However, the history is often not seen this way because the authoritative interpreters of the scientific method for roughly the last 150 years have normally constituted themselves as a professional scientific community, not a neutral judiciary. Indeed, given Bacon's generally derogatory attitude towards the Scholastics, it is unlikely that he would have welcomed the guild-like scientific disciplines that have captured control of science in the modern period. However, in Bacon's eyes, one redeeming feature of science's institutionalization over the past two centuries would be the role of university teaching in dissipating the epistemic advantage accrued by academics steeped in original research or years of deep study.

This institutional innovation – associated with Humboldt – was specifically designed to enable a new generation of enquirers to enter a discipline at a relatively level playing field by forcing expert practitioners to publicly justify (in the classroom) how their own work follows from pedagogically tractable 'first principles' in their discipline. I have likened this process to the Schumpeterian one of 'creative destruction' (Fuller 2009: chap. 1). In more contemporary terms, we might think of the Humboldtian emphasis on bringing research and scholarship into the classroom as a periodic rebooting of the academy's epistemic mainframe. It enforces a sense of *temporal democracy*, so that being born later does not constitute a structural disadvantage. In the past, this had been handled either by each generation simply being taught to repeat the classics of the past (as in the ancient Chinese civil service vis-à-vis the Confucian corpus) or by one simply possessing the time needed to recapitulate the historical trajectory of the relevant field of enquiry at one's leisure before making an original contribution (e.g. Darwin). The one strategy arrested epistemic progress entirely, the other rendered it an accident of inherited privilege.

In contrast to these counterproductive means of advancing knowledge, the requirement that new insights be test-driven on a student audience provides a Baconian 'crucial experiment' for what – with a nod to the great post-war French political theorist Bertrand de Jouvenel – might be called their *futuribility*, which is the specifically temporal version of what the philosopher of science Nelson Goodman (1955) originally called 'projectibility', namely, a wheat-and-chaff exercise that considers which features of today's knowledge are worth taking forward to serve as the starting point for the next generation – as opposed to mere artefacts of how the knowledge was originally discovered or is currently promoted. Thus, the 'futurible' may be seen as tracking truth in time.

While teaching continues to perform as a Baconian filter, at least in universities still committed to the Humboldtian ideal, the rest of the Baconian state-science settlement is under increasing criticism in our age of Protscience. Scientific authority tends to be wielded in institutions that are unaccountable to those whom they would govern. I include here national academies of science and academic journals that marginalize, if not ignore, the views of the people whose lives would be regulated, while at the same time expecting automatic deference to their authority. It is worth stressing that this point applies, in the first instance, to the *scientists themselves* and only secondarily to the general public. As scientifically accredited advocates of homeopathy and intelligent design theory can all too easily testify, those who take an agreed body of scientific data in a theoretically proscribed direction are dismissed on exactly the same terms as someone without any specialist training who happened upon similar views on the internet: that is, conformity is the primary marker of competence. This is perhaps the best evidence that Kuhn's (1970) authoritarian paradigm-driven vision of science continues to rule. Protscience aims to rejig the balance of epistemic power, so that researchers can draw significantly different conclusions from facts that are agreed by their field's orthodoxy, and doctors can treat their patients as clients who need to be sold a proposed treatment rather than be treated machines simply in need of repair.

In final analysis, the emergence of Protscience is the product of the dialectical tension inherent in the very idea of 'science democratized' that is regularly reproduced in the university's mode of knowledge production.

On the one hand, the university empowers students to decide for themselves what to believe and how to act in the world. Moreover, these powers have become greatly enhanced by the information and communication infrastructure in which more and more of social life is conducted. Thus, individuals happily invest and even risk their own personal resources as front-line knowledge producers. Here the role of what has been called the 'co-curriculum' as the seedbed of Protscience comes into full view: it is not by accident that Microsoft, Google, Facebook and *Wikipedia* were all invented as extracurricular activities of university students. It may be that the university is today incubating information-based entrepreneurship just as in previous generations it incubated revolutionary political cells.

On the other hand, however much welcomed this dynamism may be, it does force universities to reassert their normative control over knowledge production by systematically testing new claims to epistemic authority (be they beliefs or techniques), setting appropriate burdens of proof for Protscience challengers, and generating narrative contexts in which Protscience intellectual innovations can be understood and critically evaluated by those not directly party to them. In other words, the university will be increasingly compelled to exert governance over the market. In the past, the task was made easier because universities in most countries were ultimately agents of the state. (The United States is only a partial exception to this rule, given the mission of state-based 'land-grant universities'.) However, now the challenge is for universities to exert a similar authority without necessarily enjoying that political backing (cf. Fuller 2016a). The stakes are very high, for if you do not govern the introduction of innovations into the market, then the market will end up deskilling you. This maxim applies just as much to academics as to any other form of labour. A long-term test case is currently in progress as academics try to respond to the ascendancy of *Wikipedia*.

Interlude: *Wikipedia* – A Democratic Cure Worse Than Its Elitist Disease?

Wikipedia, the online encyclopaedia, is the most impressive collective intellectual project ever attempted – and perhaps achieved. It demands the attention and perhaps even the contribution of anyone concerned with the future of knowledge.

Wikipedia's true significance has gone largely unremarked because of the speed with which it has become a fixture in cyberspace. Since its sixth anniversary in 2007, *Wikipedia* has consistently ranked in the top ten most frequently viewed websites worldwide, now hovering between the fifth and seventh position.

Wikipedia is an encyclopaedia to which anyone with a modicum of time, articulateness and computer skills can contribute. Anyone can change any entry or add a new entry, and the results will immediately appear for all to see – and potentially contest. 'Wiki' is a Hawaiian root that was officially added to English in 2007 to signify something done quickly – in this case, changes in the collective body of knowledge. Over thirty million 'Wikipedians' have now contributed to 5.5 million entries in English alone, at a rate of 650 entries per day, with more than one million in 11 other languages, including Vietnamese. Moreover, there is a relatively large hard core of contributors: 125,000 Wikipedians have made at least five contributions in any given 30-day period.

As to be expected of a self-organizing process, the quality of articles is uneven but not uniformly bad. It all depends on the interpretations that the dominant editors make of *Wikipedia*'s editing principles, which in the end is about keeping the collaborative approach afloat (Tkacz 2015). This means that the topics favoured by the sex-starved male geeks who dominate the editorial process have been elaborated in disturbingly exquisite detail, while less alluring matters have been left to lie fallow. Nevertheless, *Nature*'s 2005 evaluation of the two encyclopaedias in terms of comparably developed scientific articles found that *Wikipedia* averaged four errors to the *Britannica*'s three. That difference has been probably narrowed since then. And more than a decade ago, Cass Sunstein could already observe that *Wikipedia* was cited four times more than the *Encyclopaedia Britannica* in US judicial decisions (Peoples 2010). *Wikipedia* currently has 50 times more entries than the largest edition of *Encyclopaedia Britannica*.

Moreover, *Wikipedia* has been trumpeted as heralding the arrival of 'Web 2.0'. Whereas 'Web 1.0' supposedly focussed on the ease with which vast amounts of information of different kinds can be stored and transmitted in cyberspace, 'Web 2.0' renders the whole process interactive, removing the final frontier separating the transmitter and the receiver of information. Yet, we have been here before – in fact, for most of human history. Indeed, as I observed when discussing 'information overload', the sharp divide between producers and consumers of knowledge began only about three hundred years ago, when book printers secured royal protection for their trade in the face of piracy in a rapidly expanding market for literary products. The legacy of their success, copyright law, continues to impede attempts to render cyberspace a free marketplace of ideas, as the work of the US lawyer Lawrence Lessig (2001) continues to testify.

From the standpoint of Protscience, 'open access' policies deal only cosmetically with the underlying political economy of cognitive access, namely, the educational entry costs required to make sense of materials now made 'freely' available (Fuller 2016a: chap. 3). At the same time, the gradual transfer of modern intellectual property law into a nominally 'free' cyberspace arguably legalizes the old piracy that such law was designed to prohibit, as one's original efforts can be easily usurped by someone capable of repackaging them as their own and ensuring its prominence in web searches. This is the latest twist in the field of 'infopreneurshp', which legal scholars had begun to recognize even before the invention of the World Wide Web, since the same principles had been already discussed in court cases that involve music sampling (e.g. Jaszi 1994).

To be sure, before 1700 there were fewer readers and writers, but they were the same people with relatively direct access to each other's work. Indeed, a much smaller, slower and more fragmented version of the Wikipedian community came into existence with the rise of the universities in twelfth- and thirteenth-century Europe. The large ornamental codices of the early Middle Ages gave way to portable 'handbooks' designed for the lighter touch of a quill pen. However, the pages of these books continued to be made of animal hide, which could be easily written over. This often made it difficult to attribute authorship because a text might consist of a copied lecture in which the copyist's comments had been inserted and then perhaps altered as the book passed to other hands. *Wikipedia* has remedied many of the technical problems that faced the medievals. Any change to an entry automatically generates a trace – a 'History' page – so that entries can be read in the spirit of what medieval scholars call a 'palimpsest', a text that has been successively overwritten. Moreover, *Wikipedia*'s 'Talk' page provides ample opportunity to discuss actual and possible changes of a given entry. And of course, Wikipedians do not need to pass around physical copies of their texts: everyone owns a virtual copy of the same text.

Most importantly, *Wikipedia*'s spirit remains deeply medieval in its content policy. *Wikipedia* content policy consists of three principles: (1) No Original Research. (2) Neutral Point of View. (3) Verifiability. They are designed for people with much reference material at their disposal but no authority to evaluate a knowledge claim beyond arguing from what is contained in that material. Such was the epistemic position of the Middle Ages, which presumed all humans to be mutually equal but subordinate to an inscrutable God. That too was a period that did not recognize personal expertise, only verifiable sources. The most one could hope for, then, was a perfectly balanced dialectic. In the Middle Ages this attitude spawned scholastic disputation. In cyberspace the same practice, often dismissed as 'trolling', remains the backbone of *Wikipedia*'s quality control.

Wikipedia embodies medievalism democratized. *Wikipedia*'s founder, Jimmy Wales, has justified the three policy principles on his own inability to tell whether fringe theories about, say, physics or history are true, which means that he can do no more than trace the source for a knowledge claim and its level of support, both backed by hyperlinks. Moreover, Wales's peculiar sense of epistemic suspension seems to have anchored – some might say biased – the path taken by online collective knowledge production more generally. Thus, the attempt by *Wikipedia*'s original employee, the philosopher Larry Sanger, to set up the competing 'Citizendium' encyclopaedia organized around domain expertise has largely failed, and the prospects for Wales's own anti-fake news service, 'Wikitribune', do not look especially promising.

If we take Wales's suspended judgement as *Wikipedia*'s epistemic norm, it would seem to follow that the optimal level of academic achievement for engaging in *Wikipedia*'s editing process is that of the undergraduate student. After all, the prescribed norms of conduct of students correspond exactly to *Wikipedia*'s content policy: one is not expected to do original research but to know where it is and how to argue about it. Compulsory student participation would not only improve *Wikipedia*'s already impressive collective knowledge base but also might help curb the elitist pretensions of researchers in the global knowledge system. Indeed, participation in *Wikipedia* might be made compulsory for advanced undergraduates worldwide. While this may only modestly raise the content quality of the average article, it will significantly improve *Wikipedia*'s curatorial standards, which is ultimately the source of whatever epistemic authority *Wikipedia* enjoys. Or as I say to my own students, For any given *Wikipedia* entry, take the links more seriously than the text.

For all its undisputed virtues of access, *Wikipedia* is the sort of repository of human knowledge for which the concept of a 'second opinion' might have been invented. It is to *Wikipedia*'s credit that it contains 'Talk' pages which recount – in lugubrious if not gory detail – the second, third and more opinions on offer in response to various entry edits. It is probably the closest that historians could ever get to seeing the global hive mind fitfully approximating a state of reflection. Nevertheless, the results can seem puzzling, especially when judged against *Wikipedia*'s expressed tripartite editorial policy of neutrality, verifiability and no original research.

Consider the rather detailed *Wikipedia* entry on 'Pseudoscience', which branches out into a 'List of topics characterized as pseudoscience'. Were it not for the entry's surface incoherence, which tends to be a feature of longer *Wikipedia* entries, one would never guess that 'pseudoscience' is largely a rhetorical device. It is deployed rarely by scientists, sometimes by philosophers and most often by self-appointed popular defenders of science. Thus, one finds the opinions of people and periodicals associated with the Southern California-based 'Skeptics Society' carrying significant weight in Wikipedian discussions of pseudoscience. Nevertheless, upon turning to the Talk page, one discovers that 'Pseudoscience' is classed as a 'vital article in Philosophy', which will come as news to professional philosophers of science, many – if not most – of whom believe that it is a pseudo-topic. Even the authoritative *Stanford Encyclopedia of Philosophy*'s entry on 'Pseudoscience', cited by *Wikipedia*, is quick to deflect the issue to the more generic epistemological problem of the warrant for belief, all the while putting on a brave face as it admits the persistent elusiveness of the named quarry.

As for the entry enumerating alleged pseudosciences, all the usual suspects, past and present, are included – with extra loving attention paid to psychology and medicine. Yet, three likely suspects are conspicuously absent from this otherwise thorough line-up: Social Darwinism, eugenics and sociobiology. Instead the entry makes a point of deeming 'pseudoscientific' non-Darwinian versions of social evolution that were common from the eighteenth to the twentieth century. Even when one turns to the elaborate, useful and largely sympathetic entry on 'Eugenics', the only mention of 'pseudoscience' is in a rather apologetic vein, namely, that some people think that 'improving the human stock' is not a scientific idea. One would never guess that eugenics and other extensions of 'selectionist' thinking into the human domain such as sociobiology had been considered paradigm cases of pseudoscience in the 1970s when I first came across the idea of 'pseudoscience' as a student.

As it turns out, my own views on eugenics are rather similar to those of the *Wikipedia* editors, and it should already be clear that I do not hold much store by the very idea of pseudoscience. Nevertheless, it is striking that eugenics gets off so lightly, considering *Wikipedia*'s preoccupation with pseudoscience. My guess is that a difference in generational sensibility – that is, distance from the Second World War – is at play here. People schooled in my generation would have been much quicker to ferret out insidious motives and intellectual fraud in human-oriented biological research, and some would go beyond 'pseudoscience' to call its practitioners 'crypto-Nazis'. At least, it is surprising that these concerns do not arise among the *Wikipedia* editors, whereas they would likely surface in any semi-informed conversation.

But Wikipedians are by no means indolent. In fact, the diligence previously deployed on selectionist topics are now used to ferret out insidious motives and intellectual fraud in movements such as intelligent design which try to introduce 'supernatural' considerations into the conduct and interpretation of science. To be sure, creationism has been always a staple of pseudoscience, but mainly because it claimed that the Bible was the ultimate source of epistemic authority, trumping even the latest science. In contrast, intelligent design theory aims to reintroduce biblical thinking by using the tools of science against received scientific opinion.

Interestingly, when I first learned about pseudoscience in the 1970s, establishment thinking – not to be confused with the 'relativists', 'postmodernists' and 'New Agers' routinely demonized in *Wikipedia* – was quite receptive to the blending of science and religion in aid of some synthetic 'humanist' future. Two of the originators of the neo-Darwinian synthesis in biology, Theodosius Dobzhansky and Julian Huxley, had promoted the works of the heretical Jesuit palaeontologist Teilhard de Chardin in just this spirit. However, all of this happened before a 'religious right' gained ascendancy, first in Christian America but now equally seen in terms of Islamic militancy. At that point, religion became the default enemy of science, unless its knowledge claims were clearly separated from those of science. On this specific point, the two major public antagonists over the interpretation of evolutionary theory in the closing years of the twentieth century, Richard Dawkins and Stephen Jay Gould, were in agreement.

Wikipedia editorial sensibility seems to have been formed by this relatively recent turn of events. When combined with *Wikipedia*'s implementation of its own norms, it can be difficult for readers to get a fair sense of the issues surrounding a hotly contested case of 'pseudoscience' such as intelligent design theory simply by focusing on the text of the entry page. One always needs to go to the cited sources, the Talk pages and then judge for oneself. As someone active in the intelligent design controversy, I can offer my own *Wikipedia* entry as evidence.

Wikipedia's norms provide grounds for permissible editing in the minimal sense. Thus, an entry may be written in a neutral tone, with verifiable sources and no original research, yet without representing the full range of opinion on a topic. If there is a missing opinion, it is presumed that someone will eventually provide it – and that often happens, eventually. Moreover, there is no specific commitment to providing a clear statement of a position being criticized, and certainly not in any specific proportion to the criticism published. Again, it is left for a criticized party's defenders to make their presence felt. In practice, this means for better or worse that the treatment of controversial topics can turn out to be an editorial war of attrition, which perhaps suits a medium that does not regard Social Darwinism as pseudoscience! But more seriously, it reflects the hidden hand of Hayek (1945), the source of Wales's original inspiration for *Wikipedia* (Schiff 2006).

Historical Precedents and Future Prospects for an Adequate Scientific Response to Customized Science

It would be a mistake to think that the rise of customized science is without precedent. When the state has not been the dominant shareholder in science, scientists have had to sell their work. It is no accident that public engagement with science is probably more developed in the United Kingdom than in any other scientifically advanced nation, with the Royal Institution – founded in 1799 and home to Humphry Davy and Michael Faraday – leading the way (Knight 2006). It is a historical legacy of the state's hands-off policy to the conduct of science in response to the Royal Society's chartered promise not to meddle in matters of state. Moreover, compared with other scientifically advanced nations, British scientists only relatively recently came to rely on a steady stream of state funding – which is now 'consolidating', if not drying up. The result is a research culture that is used to 'singing for its supper'. Since the nineteenth century, this imperative has been especially felt by those for whom science has been a vehicle of upward social mobility, perhaps most famously Faraday and Thomas Henry Huxley, the two poor boys who still set the gold standard for science communication in, respectively, its demonstrative and argumentative modes. In this vein, until the end of the Cold War, science was probably sold more as a secular religion – with the likes of Faraday and Huxley functioning as celebrants – than a species of venture capitalism, as it is increasingly sold today.

However, the market for science began to take a more business-like turn once the costs of doing science – ranging from the human and material entry costs to more downstream effects on society and the environment – had got so high that science had greater need for investors and stakeholders than outright practitioners. This shift began in earnest – that is, across all fields of science – with the end of the Cold War. At that point, science was thrown open to an unprotected market environment, in which science's 'value for money' could not be taken for granted. In this respect, the Cold War was the golden age for science policy because all sides agreed that science was necessary for the future of our survival – in terms of securing the physical spaces in which we conduct our lives. The threat of nuclear holocaust kept the global mind focused on the value of science. Once that threat was thought to have been removed, science had to be sold to various constituencies, each on its own terms. Unsurprisingly, perhaps, philosophers have followed the money during this transition, and hence the unified vision of physics has yielded to biology's pluralism as science's paradigmatic disciplinary formation (e.g. Dupré 1993).

The upshot is that science needs to devote an increasing amount of its own resources to proactive marketing, or *pro-marketing*. It is the third of three phases in science-led initiatives relating to the 'public understanding of science' that have occurred in the aftermath of the Cold War. The three phases are as follows:

1. In keeping with the 'deficit model' discussed above, in the final decade of the twentieth century, scientists were urged to do their own press releases to

ensure that the public is given a clear sense of their work without what scientists regard as 'journalistic misrepresentation'. This practice is still promoted in a more sophisticated form – and even rewarded (e.g. the recent knighthood of Fiona Fox, head of London's Science Media Centre) – but is no longer seen as the dominant solution.

2. At the dawn of the current century, public understanding of science took a radically prospective turn, which often goes by the name of 'anticipatory governance'. The US National Science Foundation (and later the EU) hired science and technology studies researchers to conduct market research on what people hoped and feared from what the NSF was promoting as an imminent 'convergence' of nano-, bio-, info- and cognosciences and technologies (Barben et al. 2008). The scenarios presented in the focus groups and wiki-media were speculative, but the responses provided valuable information about how to present such developments so as not to alienate the public. From a social psychological standpoint, these exercises also served to immunize the public against any 'future shock', given that discoveries tend to happen rather unexpectedly. Today's science fiction scenario may turn out to be science fact tomorrow – and one would not wish a public backlash based on what George W. Bush's bioethics tsar, Leon Kass (1997), euphemistically called 'the wisdom of repugnance'.

3. But in the emerging world of science pro-marketing, one should not merely create receptive publics for new science and technology but also make people *want* to see the innovations as integral to their own self-development. The precedent for such proactive marketing comes from the great psychologist of self-actualization, Abraham Maslow (1988), who towards the end of his life in the late 1960s proposed 'Theory Z', which encouraged people to associate their individuality from, if not superiority to, others in terms of consumption patterns based on a sophisticated knowledge of differences between goods that prima facie may not seem so different. When people fuss over whether their food has been genetically modified or their clothes were manufactured in third world sweatshops, Theory Z is in effect. The consumption patterns of such people are, as Thorstein Veblen might say, 'conspicuous' – but in this case, not to show off how rich they are but how clever they are. (Maslow's euphemistic name for this class was 'transcenders'.) Of course, in the long term, these people may be shown to have been fools for having paid more for goods based on a false vision of how the world works, but in the meanwhile their expenditure will have served to push that vision – not to mention keep the capitalist economy ticking over.

The problem to which Maslow's Theory Z provides science with a pro-marketing solution is how to increase the public's personal and material investment in science without necessarily expecting them to become – or even agree with – professional scientists. In short: *how can science build its customer base?* Even today, it is common to measure the impact of public understanding of science campaigns by the number of new recruits to science degree programmes, despite that many if not all sciences – physics most notably – would be better served by fewer recruits but more funding to secure the time, space and materials needed to settle long-standing theoretical questions for which there are now a surfeit of alternative models (Smolin 2006). To this we might add, perhaps causing more distress to professional scientists, the need for people to integrate science into their daily lives, including 'metaphorical' extensions of core scientific concepts and findings. In the history of modern market research, Maslow is credited with showing how seemingly other-worldly 'New Age' types with few traditional commitments but much disposable income and highly discriminating tastes could be a steady profit-maker for business – a latter-day descendant of which is the 'long tail' niche marketing strategy (Anderson 2006). Perhaps now it is time for science itself to cash in, even if that means cultivating some of the very people who would normally make them cringe. What follows is a proposal in this spirit.

Consider that populist successor of Carl Sagan who is now the telegenic face of UK cosmology, Brian Cox, some of whose million – now three million – Twitter followers had tried unsuccessfully a few years ago to swell the physics degree programmes at his home base, the University of Manchester. When Cox is not doing a film shoot or researching at CERN, he actively lobbies for more physics funding (Jeffries 2011). But these pursuits need not remain distinct. Cox flirts with New Age themes on television, such as alluding to astrology's early formative role in getting people to imagine that things happening in remote times and places might directly bear on who and what they are – the basis for science as a quest for the 'grand unified theory of everything'. In that case, why not team up with the San Diego-based bestselling physician Deepak Chopra (1989), who promotes 'quantum medicine' as a personalized version of this general vision? To be sure, Chopra has been denounced for practicing what the physicist Richard Feynman (1974) originally called 'cargo cult science', an allusion to the natives of Southern Pacific Islands who during and after the Second World War built life-size cardboard replicas of the airplanes that brought them food and supplies from the United States and Japan, purportedly to keep the planes coming. By extension, advocates of 'quantum healing' are equally deluded to think that by enthusing – or simply talking – about quantum mechanics, their health will be improved, as if insights from that field of physics had direct implications for medicine.

Stated so baldly, of course, knowledge claims made on behalf of quantum healing look very dubious. However, with some hermeneutical charity, one can see an indirect route to the sorts of connections that Chopra wishes to make between physics and medicine through, say, the 'quantum decoherence' theory that the mathematical physicist Roger Penrose (1989) has proposed which would effectively explain consciousness as quantum effects that are made possible by the size and structure of neural pathways in the brain. While this theory remains quite speculative, it has attracted the attention of other professional scientists interested in the prospects for spiritual life within the parameters of contemporary physical cosmology (e.g. Kauffman 2008: chap. 13).

Science customization encourages just this sort of unconventional theory construction, the end result of which may be to get the supporters of Cox and Chopra to see themselves as much more joined in common cause than they might first suppose. But such moves will only happen once more conventional supporters of science prioritize *promoting* science over simply *protecting* it. As for Chopra himself, notwithstanding the severe criticisms that his work has endured over the past quarter-century, he has attracted scientifically respectable co-authors, perhaps most notably Rudolph Tanzi, the famed Harvard-based neurologist specializing in Alzheimer's disease, to promote the hidden powers of the brain (Chopra and Tanzi 2012).

In a world where even scientists – as well as non-scientists – are in the first instance customers rather than manufacturers of science, the appropriate attitude to have to the 'knowledge' presented is one of a *shopper*. Shopping is not an activity that is normally accorded much epistemic respect, but the skills involved amount to a radically democratized version of connoisseurship. Shopping cannot be reduced to 'following fashion' because shopping implies thinking in terms of the personal suitability of a good on sale. Indeed, we tend to underestimate the amount of kickback that ordinary shoppers give to fashion. You try on the garment or test-drive the car before purchase. And even if you buy the good, you then shape it at least as much as it shapes you. 'Science customization' is just one more step in that direction.

In British English, 'shopping' has an interesting meaning that accentuates the sense of 'shopping' I am advocating. If you 'shop someone', you are informing on them to the authorities, presumably because you managed to gain their confidence, which led them to confess something criminal. This is akin to the profound Italian adage *traduttore traditore*, which I first learned in my student days when Jacques Derrida was fashionable: 'To translate is to betray'. While it is often presented as a counsel of despair against the prospect of adequate translation, the adage is best understood as a formula for turning a double negative into a positive: you give the impression that your time spent on studying some technical subject or someone's thought is to follow it, but in practice your intention is to supersede it, either by opposing it outright or creatively misappropriating it (Bloom 1973). Once you entertain this treacherous frame of mind, you are equipped to live in the post-truth condition.

Chapter 6

THE PERFORMANCE OF POLITICS AND SCIENCE ON THE PLAYING FIELD OF TIME

The Weberian Dialectic: Where Political Philosophy and Philosophy of Science Meet

The idea that science and politics are somehow metaphysically different has helped keep politically oriented academic disciplines, such as political science and international relations, distinct from the actual politics. Max Weber had an especially influential way of making the point a century ago in a couple of lectures delivered to university students in Munich: 'Science as a Vocation' (1917) and 'Politics as a Vocation' (1919). Weber saw science as *wertrational* ('value-rational') and politics as *zweckrational* ('ends-rational') pursuits. To be sure, he claimed to be talking about what he called 'ideal types' of the scientist and the politician, but for us to call them 'stereotypes' would not be inappropriate.

The scientist is principled in her pursuit of the truth without necessarily knowing the end. She is a 'realist' in that peculiar post-Kantian sense of holding herself accountable to a standard over which she ultimately has no control. This sense of a 'mind-independent' reality is the secular residue of the transcendent Abrahamic deity. Thus, the ultimate truth of our knowledge claims is akin to the Final Judgement that God passes over our lives. Method in science functions as moral codes do in such religions – not as foolproof formulas to salvation but as heuristics whose value is always demonstrated indirectly. (No surprise perhaps that the person who coined 'heuristics' – who also coined 'scientist' to name a profession – was William Whewell, the nineteenth century's exemplar of the hybrid scientist-theologian.) Popper's philosophy of science is based very clearly on this idea – that a positive outcome to an experiment does not outright confirm a hypothesis but simply fails to falsify it. Thus, the scientist is licenced to continue promoting the hypothesis, which in the long run may only have provided the scientist enough rope with which to hang herself.

In contrast, the politician is focused on 'the ends justifies the means', which can leave observers with the impression that the politician is unprincipled, even unscrupulous, in her dealings with others. However, the politician wishes to be seen as so convinced of the rightness of his vision that he will do whatever it takes to bring it about. The sheer expression of that conviction should attract enough followers to turn the vision into a reality, perhaps even in the manner of a self-fulfilling prophecy. This is quite recognizably an 'antirealist' position in the sense that was associated with 'constructivism' and 'decisionism' across a wide range of philosophical specialities from mathematical logic to legal theory in the twentieth century, all of which are concerned with the nature of *normativity* (Turner 2010). Unlike the scientist, who aims to provide a perspicuous representation of a reality that continues to exist even if she fails to represent it properly, the politician actively participates in producing the reality he wants, indeed one which he would like others to think would not have come about without her intervention. This modus operandi captures what Weber christened as 'charisma' in politics.

What Weber identified here were not two completely separate world views but two orthogonal ways of viewing the same world. In other words, 'realism' and 'antirealism' should not be seen as contradictory positions but as the same position looked at from two different angles. Here I take my cue from the late Oxford metaphysician Michael Dummett (1978), who famously organized his philosophy around the idea that realism and antirealism differ over the 'determinacy' of truth and falsehood. Realists hold that there is a fact of the matter as to whether something is true or false, regardless of whether we know it. In that sense, a statement is always 'determinately' true or false. In contrast, antirealists deny that there is such a fact of the matter until we know it – or at least until we have a procedure that concludes with our knowing it. In that sense, a statement is 'indeterminately' true or false unless we have some way of settling the matter.

Perhaps the most intuitive way of characterizing this difference in perspective – certainly one that appealed to both the logical positivists and the Popperians – is to say that realists start with the existence of a semantically closed language, in which each grammatical sentence is always already either true or false, the fact of which is determined by correspondence to a reality outside the given language. In contrast, antirealists start at the logically prior stage of having to decide which language to use. This 'metalinguistic' standpoint is ultimately a matter of 'convention', implying a free choice in terms of which way the world is to be semantically divided, and it is only once that decision is taken that the distinction between language and reality is in force. An updated version of this contrast for a generation reared on *The Matrix* appears in the media theorist Douglas Rushkoff's (2010) exhortation 'Program or Be Programmed!' The former option captures the antirealist and the latter the realist sensibility.

Kuhn (1970) notoriously qualified the realist sensibility in a way that specifically applied to science, one which appealed more to the positivists (who published Kuhn's book as the final instalment of their encyclopaedia) than to the Popperians (Fuller 2000: chap. 6). The realism of science depends not only on scientists settling on a theoretical language to which they agree to be held accountable, but also that they all settle on *the same* such language. This speaks to the authoritarian character of the 'paradigm', which is underwritten by a regime of standardized training and peer-review judgement. Here Kuhn was reflecting on the relatively short period required for the scientific community to rally around Newton's world system. This had been hastened by the Charter of the Royal Society of London, whose prohibition of matters relating to politics, religion and morals was designed to minimize the often lethal tumult that had been unleashed by the Protestant Reformation, during which the so-called Scientific Revolution transpired. (In terms of the previous paragraph, the Reformation constituted the ultimate 'metalinguistic' struggle.) In contrast, the Popperians regarded Kuhn's insight as an overreaction, preferring many 'research programmes' – their rhetorically scaled-down version of 'paradigms' – to bloom as long as each conducted itself in a methodologically rigorous fashion, as epitomized by the 'falsifiability principle'. In that case, external observers can draw their own conclusions with regard to their investments, allegiances and actions based on the track records of the various research programmes.

Latter-day analytic philosophers, who tend to be more scholastic than the Popperians or even the positivists ever were, have found it difficult to classify these precursors as realists or antirealists, as they tended to switch back and forth in perspective. This is epitomized in the ambiguous role played by *hypothesis* in both positivist and Popperian thought: 'Hypothesis' stands at once for a freely chosen principle to orient scientific inquiry (antirealist) and a testable claim about a reality that lies outside of inquiry (realist). Here one should not underestimate the significance of Gestalt psychology – not least the 'Gestalt switch' – in orienting this entire way of thinking. (Popper himself had been a student of Karl Bühler, one of the early Gestalt psychologists.) In effect, fallibility is the flip side of freedom, and in this way realism and antirealism are joined at the hip (Fuller 2015: chap. 4).

But beyond these matters of philosophical self-positioning, the difference in perspective represented by realism and antirealism also helps explain an important difference in the modus operandi of scientists and politicians – namely, the rigour of the former and the flexibility of the latter. From a logical point of view, politicians stand at a metalevel to scientists, which begins to give meaning to Bismarck's definition of politics as the art of the possible. Put another way, if you control the frame, you control the game. In the 1860s, this was described as the 'room to manoeuvre' (*Spielraum*), which the savvy politician tries to expand at the expense of opponents. By the 1960s, one would speak of 'the name of the game'. However, the consolation prize for the losers may be science, which aims to uncover the rules of the game, presumably in the hope that one may become a *magister ludi* in the future. This would amount to leveraging science to acquire political competence, something that taking the red pill is designed to achieve for the people living in the simulated world on which *The Matrix* is premised.

On Plato's telling, one would need to spend several decades in his Academy to acquire a comparable competence. But for the lawyer Francis Bacon, the generally acknowledged founder of the modern scientific method, this knowledge may be acquired by the 'experimental' study of nature, which he likened to an inquisitor's treatment of a hostile witness: both nature and the witness need to submit to abnormal conditions ('extreme experience' is close to the original meaning of 'experiment'), since neither is inclined to reveal her secrets easily, as that would remove whatever power she – and Bacon saw both the witness and nature in feminine terms – has over her investigator (Fuller 2017a). The basic idea is that different forms of torture might reveal different responses, which leaves it up to the inquisitor to determine the truth. Of course, in strictly theological terms, both 'science' and 'politics' in this sense are pursuits that easily court blasphemy for their god-like aspirations. On the one hand, scientists follow in Bacon's footsteps in their aggressive pursuit of God's exact identity by stripping away the deity's earthly guise as 'Nature', while on the other politicians such as Bismarck more simply aim to approximate God's capacity to conjure with alternative courses of action, the decision among which ideally leaves their opponents at bay.

Yet, even in a more democratized political environment, *Spielraum* reigns supreme. Popper (1957) brought the idea down to earth as the 'logic of the situation', while Weber tried to forge a concept of 'objective possibility' as the scientific correlate to *Spielraum*, in the name of his fabled *Verstehen* (Turner and Factor 1994: chap. 6; Neumann 2006). Implied here is the idea of Realpolitik, which conceptualizes politics in the business of reality construction, a competitive field in which possibilities expand and contract as an emergent effect of the actions taken by the relevant players (Bew 2016). The name of the game, then, is getting the opponent to play by your rules, so as to increase your own room to manoeuvre. In the modern world, the party-based struggles that characterize modern parliamentary democracies come closest to formalizing this modus operandi within nation states. Internationally the comparable field of play is more ambiguously defined, but 'balance of power' among nations – a strategy actively pursued by Bismarck – captures the sense of equilibrium towards which this essentially anarchic situation ideally gravitates.

Modal Power and the Fine Art of Actualizing the Possible

At stake in both politics and science – and central to the idea of *Spielraum* – is what I have called *modal power* (Fuller 2017b). Modal power consists in the capacity to decide what is and is not possible. It is the basis for the philosopher-king's authority in Plato's *Republic*. He possesses the alchemy that turns politics into science by converting his own will into a law that is binding on others and perhaps even himself. Weberian charisma is often crucial to make this alchemy work. What from the philosopher-king's standpoint is only one among many possibilities that he could have enacted becomes a necessary condition for the action of his subjects. It was for this reason that playwrights, who conjure up alternative possible worlds for entertainment, are enemy number one in the Platonic polity, as a well-acted performance can leave audiences confused about what is and is not permitted in their society. Thus, partly to assuage his teacher's concerns, Aristotle influentially argued that a well-made drama must resolve all the plot elements, thereby clearly signalling that what the audience had witnessed onstage was a pure fiction that would not be continued outside of the theatre.

Aristotle went further. He invented the concept of contingency (*endechome-non*) to capture the idea that claims about the future are neither true nor false before the fact, but will become true or false, depending on what happens. However, this concept, which aims to be faithful to how we experience the future, occludes the question of who controls the scope of the possible, in terms of which something turns out to happen or not happen. It would seem that the intellectual price that Aristotle was willing to pay to enforce a strong fact/fiction distinction was to abandon the idea of responsible power by portraying the future as inherently indeterminate in a way that the past is not. Thus, his conception of the possible ultimately resorts to a pre-agential notion of potency (*dynamos*), the basis of both the modern concept of energy and its removal from the realm of responsibility.

An early opponent of Aristotle on this point was the Alexandrian philosopher Diodorus Cronus, who presented himself as a more faithful follower of Plato and is now regarded as a proto-Stoic and an ancient precursor of modal logic. He argued that the future is either impossible or necessary, given that the future seems indeterminate only because we do not know whether it will play by the same rules of the game that we currently do. Thus, a vision of the future may appear impossible if we do not know the rules that would make it possible, while that same vision may appear necessary if we think we do know the rules. Diodorus assumed that the difference between these two starkly contrasting judgements of the future depends on whether we think the rules of the game will remain *constant* over time. However, the 'necessary' judgement may be based on our thinking that we know that a *specifically different* set of rules will be in effect – and hence we commit to play by them in advance of their formal ratification.

This risky modal strategy, which lay behind, say, Pascal's Wager for the existence of God and the self-fulfilling prophecy, is 'performative', in the broad sense that has become popular in the wake of various creative extensions of J. L. Austin's speech act theory over the past 30 years, from Judith Butler on gender to Michel Callon on the economy. In all these rather different cases, one acts 'as if' some desired regime is already in place so that it might come into place. Austin (1962) believed that this capacity to convert the possible into the actual was inherent in the semantics of natural languages. His own examples tended to come from quasi-legal contexts, such as promising, in which an entire moral regime is brought into existence through a single utterance.

The theological benchmark for this way of understanding the power of language is, of course, the Abrahamic conception of the deity who creates by pronouncing things into being (*logos*). But it can also be found in the various philosophical conceptions of 'self-legislation', from the Stoics to Kant. Here the actualizers of the possible are self-consciously finite beings who possess a moral psychology whereby one must remain steadfast in the face of a recalcitrant environment. 'Perseverance', a word favoured by Baruch de Spinoza, Hobbes and the Puritan founders of America, covered this attitude, but nowadays 'resilience' is the word of choice. All these cases preserve, in increasingly secular guise, the original sense of 'belief' as implying unconditional loyalty, which remains in the Christian sense of 'faith', itself derived from the Latin *fides*, the word used to capture the appropriate attitude of the soldier to the commander in the Roman army (Fuller 1988: chap. 2).

Jon Elster (1979, 2000) has interestingly framed this entire orientation to the world in broadly utilitarian terms as 'precommitment', whereby one freely decides to act as if the world were governed in some alternative way in order to receive the corresponding benefits. This was arguably Galileo's strategy when he made evidential claims on the basis of the telescope, even though the methodology for assessing telescopic observations had yet to be agreed. Thus, at the time of his papal inquisition, Galileo was fairly seen as a prevaricator (Feyerabend 1975). While Galileo was presuming (correctly) that the optics of the telescope would be eventually validated, his particular telescope was at best a pimped toy whose enhanced powers were based on no more than a speculative understanding of the gadget. Not surprisingly, Galileo failed to impress his inquisitors on the terms on which he was offering his knowledge claims. Nevertheless, his actions served to inspire others to play by his pre-sumed rules – and so we say he won that argument post mortem. For this to happen, both the craft and the optics of the telescope had to be developed so as to open up the horizon of possibilities which Galileo had adumbrated.

The difference between Aristotle and Diodorus that I earlier raised high-lights a more general feature in the history of humanity's attempts to come to grips with rationality in both its political and scientific guises. Perhaps the most important metaphysical difference between Aristotle's syllogistic logic and modern symbolic logic is that the former assumes that the truth values of particular statements are already known, whereas the latter – more in the spirit of Diodorus – assumes only knowledge of the conditions under which such statements might be true and what would thereby follow. The clearest way to see this is that Aristotelian syllogisms are normally expressed as a series of assertions (e.g. 'All men are mortal, Socrates is a man, Socrates is mortal'), whereas symbolic logic recasts the very same set of propositions in a hypo-thetical mode that is indifferent to the truth value of each proposition (e.g. 'If p then q, p, q').

This shift in perspective puts one in a post-truth frame of mind. It sees the actual world as just one of many possible worlds, any of which might be actionable under the right conditions. In the language of symbolic logic, the range of these possible worlds is captured in a set of algebraic equations that need to be solved simultaneously. When economists talk about 'jointly maxi-mizing' various desirable properties, this is the frame of mind that they are in. Each such equation consists of 'variables' (e.g. 'p' and 'q') that are related in terms of a 'function', which is a property that a possible world might have. In that case, the 'values' taken by the variables define that state of that world. In short, she or he who defines the terms of the equation defines out the structure of the world. Or, as the most influential analytic philosopher of the second half of the twentieth century, Willard Quine, put it, 'To be is to be the value of a variable'.

The bottom line of this perspective, which is common to both modern scientific and political rationality, is that reality is something that is decided, not given. When God decides, the result is the best possible principles for ordering the universe; when humans do it, the result is no more than a risky hypothesis that can be falsified by subsequent events. This way of seeing things is ultimately due to *theodicy*, the branch of theology concerned with explain-ing and justifying how a perfect deity could create such a seemingly imperfect world. The idea is that divine judgement is ultimately about the harmonious resolution of countervailing forces, the optimality of which is seen only upon its completion. While this occurs instantaneously in God's mind as the logic of creation (or *logos*), for humans it is extended over time, with politics and science operating as alternating horizons for understanding the process, albeit fallibly, perhaps corrigibly, but in any case with much damage done along the way. This general mode of reasoning – and all the moral qualms attached to it – is usually attributed to Leibniz, who coined 'theodicy' in 1710, only to have the very idea ridiculed as 'Panglossian' by Voltaire in *Candide*. But theodicy was soon resurrected and historicized in Hegel's 'dialectical' philosophy of history, in which from the human standpoint each moment of optimality is only temporary and indeed provides the ground for its own subsequent subversion (Elster 1978).

'As If': The Politics and Science of the Fact-Fiction Distinction

The specific 'as if' formulation of actualizing the possible, the performative expression of modal power, is due to Hans Vaihinger, whom we first encountered in Chapter 2. He built an entire philosophy around this turn of phrase (*als ob*) that Kant frequently used to discuss our attitude to reality (Vaihinger 1924). Vaihinger lived during a time when the fact-fiction distinction that Plato had done so much to emblazon in the Western mind was put under serious strain. Like Marx and Nietzsche before him, Vaihinger was strongly influenced by the demystified readings of the Bible advanced by the 'historico-critical' school of theologians who veered towards treating Jesus as more 'symbol' than deity. Two other late nineteenth-century secular trends contributed to this blurring of fact and fiction. One was the rise of 'conventionalism' in mathematics and physics, which allowed for the postulation of unprovable assumptions if they generated a logically coherent world system, which in turn might model the workings of our own. 'Non-Euclidean geometry' had been invented in just this manner, which only later was shown to provide the mathematical infrastructure for Einstein's revolution in physics. The other was the rise of the naturalistic or, as Émile Zola said, 'experimental' novel, which played out in considerable detail versions of what social reformers had imagined and sometimes witnessed to transpire in parts of society lacking any official documentation. Wolf Lepenies (1988) has shown how this development played into the early writing style of academic sociology staking out a claim 'between literature and science'.

An interesting feature in all these 'as if' cases is a general distrust of the self-certifying character of official records, whether encoded in biblical sayings, geometric axioms or national statistics. Behind the realism of the text there is always the 'irrealism' of the will that brings them into being (cf. Goodman 1978). In this context, 'irreal' should be understood in the same spirit as 'irrational' in mathematics: irrational numbers appear to exist, but they cannot be captured as a ratio of two integers, the numbers that are normally used for counting and measuring. Perhaps the most famous of such numbers is π ('pi', the ratio of a circle's circumference to its diameter). More generally, these numbers are called 'transcendental' because they cannot be exactly specified, which means that they somehow escape the normal way in which mathematical objects are produced and ordered. Considerable debate in nineteenth-century mathematics focussed on whether such numbers actually exist, with the founders of modern analytic and continental philosophy – Gottlob Frege and Edmund Husserl – playing support roles in the drama (Collins 1998: chap. 13). At stake here was the existence of a 'meta-mathematical' realm, one incommensurable with the normal range of mathematical entities but at the same time necessary – if not responsible – for the existence of those entities.

In the end, the mathematics community largely conceded that such a meta-mathematical realm was needed to explain normal mathematical entities. The insights nowadays attributed to Kurt Gödel's two 'incompleteness' theorems flow from this concession. However, Vaihinger had from the start realized that this mode of thinking has more general applicability, not least in the law. Two of the most important movements in twentieth-century jurisprudence – *legal positivism* and *legal realism* – may be understood as having picked up on complementary features of the 'as if' approach. In terms of the politics/science distinction, the former encodes the 'politics' and the latter the 'science' pole.

Legal positivism picked up on what Vaihinger called 'fictions', which he understood as pragmatically interpreting Kant's own 'transcendental' mode of philosophizing. Thus, for the legal positivist the legitimacy of particular laws rests on their derivability from what Hans Kelsen called the *Grundnorm*, which may be interpreted as the Ten Commandments, the Constitution or the social contract, depending on the unit of political authority. The *Grundnorm* is itself 'necessary' because without it none of the other laws would acquire legitimacy (Turner 2010: chap. 3). This 'antirealist' approach sees the legal system from the standpoint of the legislator who – as absolute ruler, sovereign parliament or general will – has the power to turn any pronouncement into law. In contrast, the legal realist operates from within the system and treats such legislative pronouncements and their various statutory derivations and interpretations as hypotheses to be tested against their effects on the population to which it is applied. Thus, legal realism from its early twentieth-century US roots in the 'sociological jurisprudence' of Oliver Wendell Holmes Jr and Roscoe Pound has been associated with 'judicial activism' because its adherents openly declare on occasion that certain laws simply do not work or need to be revised substantially in order to bring about policy reform in line with the Progressive movement of early twentieth-century America (White 1949: chaps. 5–7).

It is worth observing that both legal positivism and legal realism are often seen as 'revisionary' approaches to the law because practicing lawyers – including judges – do not normally register such a heightened sense of the power dynamics involved in maintaining the integrity of the law as a closed system. In this respect, these two schools of jurisprudence – notwithstanding their prima facie divergence in approach – operate within a 'post-truth' horizon, one that presumes what Paul Ricoeur (1970) famously called a 'hermeneutics of suspicion' with regard to seemingly established 'black letter' issues of the law. This again recalls the original spirit of Vaihinger's 'as if' philosophy, albeit in this case one where players on *both* sides of *The Matrix* take the red pill.

So far this discussion of the 'fictionalist' perspective on the law has occurred from the standpoint of a single state-based legal system. However, once we expand to the interstate system, do matters remain the same or differ? This has been arguably the central practical and philosophical question of international relations in the modern era, especially once it became clear that there might be good political reasons to include a broader range of people as residents and citizens of a 'state' than the nineteenth-century ideal of a 'nation state' would suggest. This recognition of the essential artificiality of the concept of 'state' led Kelsen to easily scale up his thinking to the international level, which enabled him to play a supporting role in the establishment of both the League of Nations and the United Nations. In these cases, his principal aim was to secure a neutral procedure to adjudicate competing normative claims, a higher-order *Grundnorm* that could serve as the rules of the universal game that all states were framed as implicitly playing, on the basis of which a determinate outcome could be reached for each case brought to trial. The International Court of Justice remains the legacy of this line of thought.

However, Kelsen's great contemporary nemesis was the Weimer jurist and later Nazi apologist, Carl Schmitt, who shared the same fictionalist orientation to the law but denied the legitimacy of any scaling up from intra- to interstate relations on the grounds that it would mean ceding at least a measure of state sovereignty, which invariably would be biased towards the interests of one nation or set of nations over those of others. And while it is perhaps now easier to side with Kelsen over Schmitt given their particular political choices across very long lives, it is worth observing that Schmitt's arguments largely echoed Max Weber's own original scepticism about the invocation of 'humanity' or 'human rights' in international law, which in turn has been repurposed in recent years by theorists of traditionally disenfranchised peoples who have felt 'othered' by self-declared universalist conceptions of international law (Bernstorff and Dunlap 2010).

The Quantum Nature of Modal Power

Our failure to register modal power means that we tend to have a flat-footed understanding of how history works. For example, much is made of the predictive failures of Marxism, starting with Marx's own failure to predict that the first revolution done in his name would occur not in the country with the best-organized industrial labour force (Germany) but in a country with a largely disorganized and pre-industrial labour force (Russia). Yet this way of putting matters gives the misleading impression that Marxists and their opponents were simply spectators to history, when in fact they were anything but that. Indeed, the phrases 'self-fulfilling' and 'self-defeating prophecies' were coined in the twentieth century to cover the peculiar forms of success and failure to which not only socialists but also capitalists – in terms of investor confidence in the market – have been prone in the modern era. People deliberately act to both increase and decrease the probability that specific predictions come true. The resulting phenomena are often discussed as the 'interactive' effects of 'observer' and 'observed, a distinction that after Niels Bohr, Werner Heisenberg and Erwin Schrödinger is associated with the workings of quantum reality.

The most natural way to interpret the mathematics of quantum mechanics is that it envisages reality as a possibility space, in which the actual world consists in the ubiquitous collapsing of this space into moments, which provide portals to understand what is possible in both the past and the future. These 'portals' are what we normally call the 'present', the arena in which cause and effect are most clearly played out. But as the content of the present changes, so too does our sense of what has been and will be possible. In that respect, nothing need be forever impossible because the right event could alter the possibility space decisively. But similarly, something that had been possible may subsequently become impossible. To be sure, my characterization is much too crude for a physicist. Nevertheless, even this crude account may offer insight to theorists of politics and science, at least in terms of how to conceptualize possibility and temporality, the two foundational categories of historiography. Thus, I will not delve into the mysteries of 'quantum causation' (aka action at a distance), let alone how a vision of reality that was designed to understand the smallest of events can be scaled up so easily to make sense of normal-size events and even overarching tendencies. However, leading international relations theorist Alexander Wendt (2015) has done much of the necessary spadework to theorize credibly about politics while taking the technicalities of quantum mechanics seriously.

The idea that events determine the course of history is a commonplace – albeit a contested one among philosophers of history. This idea is normally understood either in terms of a 'founding moment' or a 'turning point'. In the former, the past appears as a chaotic field, which the founders bring into some sort of lasting order; in the latter, the past is presented as a default pattern which the turning point upends and redirects. Kuhn's (1970) famous theory of scientific change combines the two as alternating phases of 'normal' and 'revolutionary' science. But common to both versions of the idea, seen as either distinct or complementary historical horizons, is that the 'stuff of history' is better captured – at least metaphorically – as transformations of matter than the reconstitution of possibility space. An example of the difference is the common-sense proscriptions against 'affecting' let alone 'changing' the past, even though we seem to have no problem talking about 'affecting' and even 'changing' the future. Our intuitions about time having a direction are grounded in this observation. In contrast, taking the 'quantum turn' in the sense promoted here would entail recognizing every event as potentially altering both the past and the future at once.

The asymmetry in our default judgements of temporality suggests that our ordinary intuitions about the nature of causation are incoherent, which in turn may reflect an excessively underdetermined conception of free will (Dummett 1978: chaps. 18–21). In other words, when we talk about 'changing the future', we imagine giving shape to something that remains unformed at the time of our action, full stop. Yet, it is only in retrospect – that is, once that 'something' has been given shape – that we can judge our action's efficacy in turning what had been a possible future into the actual present. We think we made a difference because the difference we see is one which we see ourselves as having made. (This is the problem that the law faces when trying to determine who should 'take responsibility' for action during a trial.) In short, our understandings of the past and the future are formed simultaneously. Indeed, the 'present' may be defined as the site where a 'possible future located in the past' is converted into the 'necessary ground for constructing the future'.

What applies as a principle of our own mental equilibrium extends to our judgements of history as a whole. For example, to claim that Isaac Newton and Henry Ford 'changed the course of history' presupposes a correspondence between what we take them to have wanted to achieve and what we take them to have achieved. It strikes a cognitive balance between the future they were projecting from the past and the present that we project into the future. Understood as an economic exchange, we forfeit a measure of our free will by letting Newton and Ford set the initial conditions by which we are able to act, but in return we acquire a sense of the direction of travel, in terms of which we can exercise our own free will in a way (we bet) that will be appreciated by future observers. In quantum terms, we concede position to receive momentum. The easiest way to see this point is in terms of our ability to insert

Newton or Ford into our own world by casting what we are now trying to do in terms of something they too were trying to do. That is the concession. But that concession then enables us to claim that we are doing things that they were unable to do. This is the power we receive from the concession, which turns the future into a field of realizable prospects.

A crucial feature of this arrangement is that we do *not* say that we are now doing things that Newton or Ford could not have imagined or recognized as part of some project they were pursuing. Were that the case, it would be difficult to credit them with having changed the course of *our* history. They might still be, in some sense, 'great' or 'interesting' figures – but not of 'our' world. Indeed, there are many such figures who are, so to speak, 'marooned' on the shores of history because they fail to offer us existential leverage. This is normally what we mean when we say they have been 'forgotten'. Yet these figures always remain to be appropriated to construct the basis on which we might move into the future. When Kuhn described the history of science's default self-understanding as 'Orwellian', he had something like this in mind (Kuhn 1970: 167). Put more explicitly, scientists do not normally realize how the significance of past research and researchers is routinely tweaked, if not airbrushed, to motivate current enquiries. During a 'scientific revolution', certain researchers and fields of research may be added or subtracted altogether. For historians of science this modus operandi does a gross injustice to the past, but for working scientists it is an acceptable price to pay for whatever new findings might result. It involves the sort of ruthlessness that would meet with Marxist approval, as I shall suggest below.

Prolegomena to a Quantum Historiography of Modal Power

For the past 50 years or so, it has been common for historians to enjoy the moral high ground in this particular disagreement. In other words, scientists generally understand that the versions of the history of science that are purveyed in science textbooks or popular science writings do not primarily perform the function of saying what happened in the past. In practice, the scientist cedes jurisdiction to the historian for deciding what is true or false about what those accounts say. In return, the historian refrains from pronouncing over the truth or falsehood of what scientists say about the future. To be sure, this division of labour – or cordon sanitaire – is not strictly observed, but it captures the normative expectation of the world in which we live.

In contrast, political history is much more self-consciously 'quantum', in that professional historians do not generally enjoy the same privilege in framing the terms in which claims about the past are validated. The Holocaust is an interesting exception – a major political event in which professional historical judgement rules, perhaps most dramatically in the 1996 UK court case *David Irving v. Penguin Books Ltd*. But that may be simply because no major political party finds it in its interest to capitalize on the Holocaust by linking it to events with which it wishes to be associated. Thus, the Holocaust exists as a self-contained moment surgically separated from the field of political play. Otherwise, as George Orwell declared in the 4 February 1944 edition of the UK democratic socialist magazine *Tribune*, 'History is written by the winners'. Little surprise, then, that the most self-consciously 'revolutionary' movement of the modern era, Marxism, has been always susceptible to bouts of 'historical revisionism', when attempts are made by more learned partisans to redirect the future by refocusing the past. Revisionism is perhaps most sympathetically seen as a more economical means of achieving what might normally require bloodshed, namely, what Trotsky called a 'permanent revolution'. The politics of 'permanent revolution' amounts to a quantum approach to history.

Interestingly, in his famous 1965 debate with Kuhn, Popper (1981) also spoke of his own falsifiability criterion as licensing a 'permanent revolution' in science. The analogy can be understood as follows. A common stock of knowledge can be extended in many different, even contradictory, directions, depending on which bit of it is put at risk in an experiment. Popper argued that science advances only when such risks are taken, the inevitable consequence of which is that scientists discard – or at least radically reinterpret – what they previously held to be true in order to enter the horizon of possibilities opened up by the experimental outcome. Popper always had in the back of his mind Einstein's move to interpret time not as universally constant but relative to an inertial frame of reference, given the outcome of the Michelson-Morley experiment. This move did not merely overturn Newton's hegemony in physics, but it also transformed Newton's dogged opponents over the previous two centuries, such early advocates of relational theories of time as Leibniz and Ernst Mach, from cranks and sore losers into heroic and prescient figures whose works were subsequently reread for clues as to what might follow in the wake of the Einsteinian revolution (Feuer 1974).

The difference between Kuhn and Popper on the role of revolutions in science can be summarized in terms of their contrasting approaches to time: *chronos* versus *kairos*, the two Greek words that Christian theologians sometimes use to contrast the narrative construction of the Old and the New Testament. In *chronos*, genealogical succession drives the narrative flow, with revolutions providing temporary ruptures which are quickly repaired to resume the flow. Thus, the order of the books of the Old Testament follow the order of patriarchs and dynasts. This is also the spirit in which Kuhn's historiography of science proceeds – that is, according to paradigms that generate normal science, occasionally punctuated by a self-inflicted crisis that precipitates revolution, the outcome of which serves to restore the natural order. In contrast, in *kairos*, there are recurrent figures who constitute the narrative but no default narrative flow, as the world order is potentially created anew from moment to moment. Thus, the New Testament begins four times with the varying Gospel accounts of the rupture that was Jesus, with all but the final book presenting various roughly contemporaneous directions in which Jesus's teachings were taken after his death, virtually all adumbrating more ruptures in the future. This is more in the Popperian spirit of presenting science as a sensibility that can be actualized at any moment to reconfigure all that had preceded and will succeed it.

The *chronos* approach clearly corresponds to the linear time of classical physics, and the *kairos* approach to the more ecstatic conception of time afforded by quantum physics. However, in conclusion, it is worth mentioning an in-between position, especially given its salience in the history of international relations. It is the idea of *perpetuity*, especially as understood in early modern philosophy to refer to the choice that God always has whether to continue or alter the universe from moment to moment. It was designed to get around a concern introduced by Aristotle's main Muslim interpreter, Averroes, that in creating a world governed by natural law, God forfeits his own free will. This would seem to imply that natural law exists 'eternally' without divine intervention. In contrast, the 'perpetualist' says that God actively maintains – or not – the law. As a conception of divine agency championed by the likes of Descartes, perpetuity did not survive the Newtonian revolution in physics. However, it persisted in political debates concerning human self-governance, especially with regard to the duration of any social contract that is struck between free agents. The idea of regular elections is perhaps the principal legacy of the 'perpetualist' mindset, reminding citizens that ultimately they are free to decide (collectively) whether or not to carry on with the current regime. More ambitious thinkers, not least Kant, believed that if all regimes were of this sort, then perpetualism could be scaled up as a principle of world governance, resulting in what he dubbed 'perpetual peace', one of the inspirations for the United Nations.

Chapter 7

FORECASTING: THE FUTURE AS THE POST-TRUTH PLAYGROUND

Introduction: The Alchemy of Deriving Truth from Error

In the past, under the inspiration of Popper, I have argued that fundamental to the governance of science as an 'open society' is the *right to be wrong* (Fuller 2000a: chap. 1). This is an extension of the classical republican ideal that one is truly free to speak their mind only if they can speak with impunity. In the Athenian and the Roman republics, this was made possible by the speakers – that is, the citizens – possessing independent means which allowed them to continue with their private lives even if they are voted down in a public meeting. The underlying intuition of this social arrangement, which is the epistemological basis of Mill's *On Liberty*, is that people who are free to speak their minds as individuals are most likely to reach the truth collectively. The entangled histories of politics, economics and knowledge reveal the difficulties in trying to implement this ideal. Nevertheless, in a post-truth world, this general line of thought is not merely endorsed but also intensified.

Behind the right to be wrong is a sense that humanity has made progress over its animal ancestors by our ideas dying in our stead (Fuller 2007b: 115–16). The long-term trend has been certainly towards minimizing the personal consequences of error. Consider three indicators. First, fewer people are put to death, or even tortured, for holding heretical beliefs. And when such things do happen, the violence committed – even if purportedly in the public interest – is roundly condemned. To be sure, whenever possible, heretical views are put to a strict 'scientific' test, the outcome of which may humiliate one of the parties. But that is as far as it goes. Second, environments have been made increasingly 'smart' so as to lower the likelihood that people will die or be severely harmed simply as a result of getting lost or misusing an artefact. Third, law enforcement's objective has gradually shifted from stopping to preventing crime, and even punishment has been increasingly interpreted in terms of compensation rather than retribution or even rehabilitation.

In all these cases, the propensity for error is far from outright eliminated. In some cases, it might even be encouraged. But in any case, error is made socially functional – and sometimes even rendered an instrument of some higher good, in the spirit of Popper's original imperative for scientists to falsify hypotheses whenever possible. This peculiar transvaluation of the false is a hallmark of the post-truth world. It almost makes it attractive to commit error. After all, something shown to be true merely confirms our expectations unless we had supposed the opposite to be the case. In this respect, the commission of error is more informative than the possession of truth: a minor triumph of doing over being. For this reason Popper fully endorsed Bacon's proposal for expediting human knowledge by the regular construction of 'crucial experiments', which are designed so that only one of two contesting hypotheses could be true. Yet the 'could be' reminds us that even the hypothesis that survives such a trial merely enjoys the privilege of undergoing another round of testing against future contenders to the truth.

In this chapter, I shall explore this socially structured cultivation of error in the case of *forecasting*, our knowledge of the future. The core activity of forecasting, prediction, is prone to routine and sometimes enormous error. What passes for expertise in this domain consists largely in protecting oneself from the negative consequences of error. This sounds like a cynical way of putting matters, but a post-truth orientation would suggest a more open-minded attitude to what I have just said. After all, one can only succeed in the long term if one survives in the short term – and the threshold for survival is lower than that for success. Indeed, in a fluid environment (due to changes in conditions or knowledge) being no more than 'good enough' may even be better than being 'the best' in the short term if one is interested in being 'the best' in the long term. Imagine that you are playing a game that extends over a set of matches against a variety of opponents and that your resources are limited. The last thing you want to do is to invest all your efforts in conclusively beating the first opponent as if that would take care of all the other opponents you encounter later. To be sure, you want to remain competitive, and so a win against the first opponent would be desirable. But a win does not require destroying the opponent. Even a draw might be acceptable under certain circumstances. In any case, you want to remain open to whatever subsequently comes your way – and not, as the evolutionists say, become 'overadapted' to a particular environment that is bound to become less salient over time.

Herbert Simon (1981) originally called this strategy of converting short-term survival into long-term success 'satisficing'. The word was coined in the context of trying to characterize the intelligence exhibited by organizations that routinely commit errors in decision-making without forfeiting their legitimacy as rational corporate agents. A key feature of such organizations is that they are in control of their narrative. They are regarded as the ultimate authority with regard to how their error-prone history is read. This applies no less to science as a collective enterprise. The influential historian and philosopher of science Thomas Kuhn (1970) attributed science's appeal with both its practitioners and the general public to its relatively uncontested and systematic interpretation of error as instrumental in the pursuit of some further truth. Controversially, Kuhn also showed that science's self-presentation as a progressive project involves a regular, 'Orwellian' rewriting of its history. In particular, unexpected findings and outcomes that would have been once treated as anomalies, if not outright errors, may come to be seen as harbingers of revolutionary breakthroughs. The ease with which liabilities can be converted to virtues in official histories of science, such that science always appears to be succeeding at getting at the truth no matter what it does, bears the hallmarks of a post-truth understanding of the world.

The next section considers the foundational epistemological issues in forecasting. These pertain to the epistemic significance of prediction and the past's relevance to the future. The rest of the chapter draws on the work of the most intellectually ambitious theorist and researcher of forecasting in our times, the US-based political psychologist Philip Tetlock. He treats forecasting as an expertise that can be evaluated and corrected scientifically. But his research also offers a blueprint for challenging expertise more generally. Tetlock gets experts to consider counterfactual scenarios ('taboo cognitions') that are far from their epistemic comfort zones, that is, prospects whose realization are deemed highly unlikely or undesirable. Tetlock classes these scenarios as 'thinking the unthinkable', recalling the title of Herman Kahn's notorious 1962 book about our prospects of surviving a limited nuclear war. This intellectual strategy has historically had explosive consequences for both knowledge and morality by extending our sense of what is feasible. More recently, Tetlock has proposed the concept of 'superforecasting' to describe forecasting as an iterated activity in relation to an overarching long-term strategy. This idea, taken from the military, helps reveal the social psychology underlying successful forecasting, which turns on the capacity to cope with – and perhaps even capitalize on – unwanted outcomes, or 'adaptive preference formation'. Taken to its logical conclusion, this strategy instils a normative stance towards the world that I have dubbed *precipitatory governance*.

Prediction: The Past's Relevance to the Future

Central to forecasting is *prediction*, which performs three epistemic functions. The first pertains to our relationship to our brains, the second to our relationship to our environment and the third to our relationship to the future.

The first epistemic function of prediction is to enhance the brain's normal operating procedure. Specifically, predictions force us to simulate a state of the actual world, specifically a future that is sufficiently distant from the present that is, strictly speaking, unknowable. This provides a qualified challenge to how we normally register the world, which is as a horizon of possibilities that are realized to varying degrees as we move through space and time, always being updated in light of external cues. This fluidity in the horizon of our prospects in turn helps explain the ease with which our memory and imagination change. Egon Brunswik, the Viennese (later Berkeley-based) psychologist who was his field's closest fellow traveller of the logical positivists, took this fluidity as the foundational insight of his discipline (Hammond and Stewart 2001). However, more recent neural network-based models of the brain have conferred on Brunswik's sense of fluidity a more explicit sense of fallibility. Through a combination of genetic front-loading and prior experience, the brain is now seen as programmed to presume that the world is more predictable than it really is. As if in vindication of Popper, what we normally call 'learning' consists mainly in error correction, a continuous process of compensating for our default sense of an overdetermined world (Clark 2016). Put another way, the mistakes that we routinely make in the sort of medium-to-long term predictions involved in forecasting help improve the brain at the very thing that it appears designed to do.

The second epistemic function of prediction is to remind us that we inhabit the world that we think about. Predictions are specific public events that do not enjoy the luxury of accommodating the full range of world views that people normally carry around in their heads. Those world views, when treated as private entertainments, are easily adjusted in light of events in which those entertaining them have no personal investment. But predictions require taking decisions in the open, which demand just such investments. Hence they can exact unwelcomed costs to one's wealth, reputation or friends, regardless of the truth of the prediction itself. The figure of Cassandra in Greek mythology springs to mind as someone who both won (in terms of accuracy) and lost (in terms of reputation) by making a prediction. More generally, those who are inclined to issue predictions – even when they predict an impending catastrophe – are revealing their own openness to personal risk. In that sense, a culture that encourages people to make predictions is inherently *proactionary*, which means that it subordinates the fate of the predictor herself to the potential benefit to society that comes from her having made the prediction, regardless of its outcome. Arguably a cultural sensibility that favours bold predictions has enabled science to make as much progress as it has, which many *precautionary* thinkers consider in retrospect to have been reckless. The proactionary/precautionary distinction will figure in the second half of this chapter (Fuller and Lipinska 2014: chap. 1).

The third epistemic function of prediction also derives from its public character. Because predictions are events experienced by predictors, their clients and interested onlookers, preparations of various sorts can be made in the run-up to a prediction's outcome, aimed at either facilitating or impeding prediction's accuracy. Although professional forecasters may be loath to admit it, predictions may even be said to license, if not authorize, these activities. Here we enter the realm of what Merton originally called 'self-fulfilling' and 'self-defeating' prophecies (1968a: chap. 13). Although Merton and others have cast these prophecies as 'pathologies' of prediction, the label fails to do justice to their rationality. After all, they involve exactly the same recognition of the co-dependency of prediction and outcome that motivates scientists to construct 'crucial experiments' and other 'test cases', which serve to force the outcome of a prediction to appear in a timely manner within certain preconceived constraints – something that probably would not have happened had the scientists not been keen to obtain a decisive result, one way or the other. Indeed, Bacon may well have been correct to see in experimental science a protocol for expediting the pace at which nature reveals its secrets to humans, namely, by the systematic priming of both humans and nature to specifically crafted outcomes.

Turning more specifically to forecasting's focus on the future, it is worth stressing that forecasting is only one way to be oriented to the future, since there are other ways which would make forecasting unadvisable, if not impossible. These other orientations would be guided by one or more of the following propositions: (1) The future is in principle unknowable, perhaps because it does not (yet) exist, and one can only know that which exists. (2) The future is already known to be different in all respects from the past and present, perhaps due a fundamental indeterminacy in nature, to which human free will contributes significantly. (3) The interaction effects between prediction and outcome are so entangled that every forecast generates unintended consequences that are bound to do more harm than good.

The above three propositions are arguably mutually reinforcing – certainly (2) and (3) are. What they share is a belief that our presence as epistemic agent is of little help – if not downright counterproductive – to our 'grasp' of the future, be it understood cognitively or practically. Much of 'therapeutic' philosophy from the Hellenistic period onward (e.g. Epicureans, Sceptics) has been an accommodation to the prospect that many – if not most – of our cherished beliefs and desires will turn out to be false or not live up to expectations. In this sort of world, forecasting is a futile activity that is bound to lead to just more personal misery. In that case, one should learn to cope with whatever happens in the future, not to try to predict, prevent or promote it in any way. Indeed, that would be the ultimate lesson that the therapeutic philosophers draw from Cassandra's fate.

So what sort of metaphysics is required for forecasting to be possible? Three basic conditions need to be in place: First, the future is sufficiently like the past and present that it makes sense to speak of 'learning' from the past and present in order to 'inform' our understanding of the future. Second, the agents responsible for bringing about and inhabiting the future are sufficiently like us that the terms of our forecast would be something recognizable (and perhaps even welcomed) to them. Third, reality is not so indeterminate that it would be impossible to tell after the fact whether – or by how much – a forecast has turned out to be right or wrong.

Of course, even granting the above three conditions, it does not follow that forecasting is easy or even likely to succeed. Arguably the success of chance-based wealth producing institutions – from insurance companies to casinos and lotteries – is testimony to people being at once both well disposed and ill-equipped to make forecasts, not least when they implicate their own fate. Much of this may be explained in terms of what social psychologists call the 'fundamental attribution error', whereby people overestimate the significance of their intentions in explaining their own success, while overestimating the significance of 'chance' in explaining the success of others (Nisbett and Ross 1980).

However, even if one takes into account the various cognitive liabilities that can affect forecasting, there remain at least three fundamentally different approaches to the dynamics of lived time that can affect the forecaster's judgements of a plausible future. These are best understood as alternative interpretations of the expression 'legacy of the past':

1. *The past provides a cumulative load on the prospects for the future through the sheer repetition of its presence over a long period.* This is the idea behind 'straight rule induction', which says we should expect the sun should rise tomorrow simply because it rose yesterday. A more sophisticated, second-order version is 'trend extrapolation', whereby we assume that the rate at which things have been changing for a long time will continue into the indefinite future. But whether we imagine that the past's 'cumulative load' is borne by our genes, our memory, our behaviour or even some invariance in nature beyond our direct control (e.g. the reliability of sunrise), the transmission process itself may allow for non-trivial changes ('mutations') over time which could decisively alter the projected outcomes.

2. *The hold of the past on our future prospects diminishes over time, unless the past is actively maintained.* Unlike point 1, here the past is not conceptualized as the original and continuing agent of history. Rather, it is treated as an increasingly distant if attractive place, any connection to which requires 'institutions' understood as indefinitely binding social arrangements whereby

successive generations of individuals occupy roles which together uphold the relevant values that were established in an ever-receding past. Against this backdrop, charges of corruption are significant because they call into question the integrity of the transmission process. In late medieval law, the term *universitas* ('corporation') was coined to capture such institutions, which originally applied to churches, universities and cities – and much later to business firms. However, the key modern application has been to the state, understood as that artificial person whom Hobbes dubbed 'Leviathan'. This entity is explicitly committed to a sense of constancy (and hence the common Greco-Latin root of 'state' and 'static' in *stasis*) over time, a sense of perpetuity that simulates on earth the sense of eternity that only God possesses.

3. *The past is a repository of possible futures only some of which are realizable at any given time.* This suggests the idea of 'paths not taken', which may be revisited in a future that in relevant respects resembles an earlier decision-making moment – yet where the circumstances are now sufficiently different to militate a different path to be followed. Presupposed here is that there are always many competing futures at any point in time, and that the one which prevails is contingent and hence potentially – if not easily – reversible. This is another angle on the idea that the present needs to be actively maintained because it will not necessarily persist into the future. But here the reason is the presence of other possibilities vying for realization. This in turn gives the status quo its authoritarian cast, and institutions appear not only to promote the status quo against degeneracy but also to suppress any alternatives from flourishing.

Does Science Help or Harm Forecasting? The Fate of Punditry as Expertise

Political psychologist Philip Tetlock is perhaps the leading contemporary student of 'expert political judgement' – that is, forecasting of the sort practiced by pundits and pollsters as well as professional politicians and policymakers (esp. Tetlock 2005, Tetlock and Gardner 2015). The epistemological upshot of his work challenges the time-honoured distinction between *political judgement* and *political science*, or more generally, the experienced practitioner and the theoretically informed scientist. To be sure, the distinction between political judgement and political science has been drawn at different times, on different grounds and to different effect. For example, Plato's 'political science' might be read as aiming to transcend the volatility of opinion in which Aristotle's conception of 'political judgement' naturally swam (and arguably rendered a virtue, given the value Aristotle placed on rhetoric). However, as an academic discipline called 'political science' emerged in the twentieth century that self-consciously modelled itself on the methods of the natural sciences, 'political judgement' became increasingly consigned to that amalgam of tradition-cum-personal experience that is often mystified as 'tacit knowledge'. In terms of this divide, Tetlock operates on the assumption that anything worthy of the name 'political judgement' is ultimately a species of applied political science. In other words, politicians and political scientists are engaged in the same sorts of fallible yet corrigible cognitive processes, albeit subject to rather different time and resource constraints.

To appreciate this perspective, consider failed predictions that are 'almost right' (but wrong) *versus* 'almost wrong' (but right). Tetlock (2005) treats the former as self-justifying defensive responses to predictive failure and the latter as the presumptive epistemic state of most predictive successes. In effect, Tetlock holds all political predictors guilty of systemic error, until proven otherwise, because by playing on the ambiguous epistemic space between 'almost right' and 'almost wrong', it is easy to leave the impression that one is exactly right. For example, one of the best publicized 'successful' social scientific predictions of recent times, namely, the collapse of the Soviet Union, was made in 1978 by the US sociologist Randall Collins (1995). On closer inspection, the prediction turns out to be both 'almost right' and 'almost wrong' but in any case, not exactly right. While Collins's general theoretical model was vindicated in that he predicted that the tipping point would come from social movements mobilizing around dissenting elites, it failed to assign a specific role to the media in expediting this mobilization. Thus, a process that on the basis of past revolutions Collins had predicted would take 30–50 years ended up happening in a little more than a decade.

Chapter 1 of Tetlock (2005) provides a set of compelling reasons for not trusting correct predictions of any sort. After all, anyone who predicts regularly is bound to hit a few targets – and if you predict the same target long enough, you are bound to hit that one. In these cases, one should be looking for the predictor's success/failure ratio. There is also the methodological problem of deciding exactly what had been predicted and whether it actually occurred. Unless one is predicting a specific outcome in an independently determined process, such as an election result, one may be faced with a hopelessly moving target, as predictors rationally reconstruct both their initial and final knowledge states. Finally, Tetlock interestingly suggests that successful predictions may be based on normatively dubious forms of knowledge that involve insider information, torture and coercion, or quite simply, models of the human condition that capitalize on deep-seated 'irrational' tendencies that we might be otherwise wish to minimize or eliminate in the future.

Tetlock's gold standard of political judgement combines the two leading philosophical theories of truth, *correspondence* and *coherence*. Thus, one predicts the right outcomes *and* for the right reasons. On the one hand, the predictor must do better than being 'right but almost wrong', in that she must have a coherent causal account of why an event should have occurred, not just the simple claim that it would occur. On the other hand, the predictor must do better than being 'wrong but almost right', in that her conceptually sound model must also capture specific actual outcomes. This is the context for understanding Tetlock's appropriation of the contrast, popularized by the Renaissance Humanist Desiderius Erasmus, of the *fox* and the *hedgehog* – the former knowing many little things and the latter one big one. Foxy predictors are always on the verge of incoherence, whilst the hedgehogs fail on strict accuracy.

The value of crystallizing political judgement as the ability to predict the right thing for the right reasons is that it forces you to confront the modal structure of your world view, not least the amount of control that you ascribe to the agents populating it. Instead of parsing the difference between hedgehogs and foxes in terms of the number of things that they 'know', one might say that foxes operate with what I have called an *underdetermined* world view, one that allows for many reversals of fate, whereas hedgehogs adopt an *overdetermined* world view that posits a dominant tendency onto which all paths ultimately converge (Fuller 2015: chap. 6).

Foxes see hidden triggering conditions to a new order where hedgehogs detect only temporary interference to the main narrative flow. Much of this boils down to historical standpoint: those for whom politics consists of lurching from one election to the next tend to behave like foxes, whereas hedgehogs are more likely to see politics as long-term movements of social transformation. Much of the authority that hedgehogs command in the media – something that Tetlock bemoans – may be attributed to their ability to speak as if their world view is destined to triumph, and hence they focus mainly on explaining how exactly that is supposed to happen. They make up for their predictive inaccuracy by personifying the future they would see come about. In that case, as we shall see, it may be wiser to follow Machiavelli than Erasmus and regard not the hedgehog but the *lion* as the species best placed to oppose the fox.

Improving political judgement in order to improve political forecasts and to improve the conduct of public life are two distinct goals. Each is difficult in its own way, the latter perhaps more so. Tetlock wants to achieve both. On the one hand, Tetlock supports the cultivation of intellectual virtue in the punditry, to improve the rigour in what he takes to be ultimately applied political science. On the other hand, would the conduct of public life really be improved if pundits were nudged to issue more accurate predictions, to own up to their failures and to update their beliefs in ways that conform to the probability axioms and Bayes' theorem? The answer is not clear, mainly due to lack of evidence about the consequences of such changes.

Consider a more modest policy to improve the conduct of public life that does not require any instruction of the punditry: a careful media record of pundits' success ratio. Even if this policy would improve public life, would it be the result of people gravitating to the pundits with better track records – or might they respond perversely? For example, a string of 'near miss' predictions by one of Tetlock's dogmatic hedgehogs might spur those resonant to her message to transform the political-economic order in ways that serve to increase the hedgehog's future predictive accuracy. In other words, false predictions can be interpreted as evidence for a systematically suppressed reality that deserves full expression – perhaps in the spirit of the third interpretation of the 'legacy of the past' in the previous section. We normally think of self-fulfilling and self-defeating prophecies as resulting from the *prior* announcement of a prediction that then triggers actions contributing to its fulfilment or defeat. But what about a *self-avenging prophecy* inspired by the attractiveness of the alternative reality inferred from a string of false predictions?

At this point, we need to add to Tetlock's political menagerie. The fox has had many foils – not only Erasmus's hedgehog but also Machiavelli's lion: The lion rules by focused shows of force, as opposed to the fox's diverse displays of cunning. Whereas the fox cajoles and adapts, often with the net effect of dissipating its energies, the lion is inclined simply to eliminate opponents who stand in its way. In that respect, the lion extends the hedgehog's explanatory idée fixe into an explicit political strategy, as, say, when a Marxist academic crosses the line from possessing a hegemonic theory to becoming a politician possessed by a hegemonic ideology.

Tetlock's failure to take the full measure of leonine politics can be seen in his conclusion that hedgehogs fare better in static environments and foxes in more dynamic ones (Tetlock 2005: 251). He presumes that both political animals simply adapt to the world – albeit each in its own inimitable way – rather than create new opportunities, let alone recover lost ones intimated in their failed predictions. This reflects Tetlock's treatment of political judgements as, to recall the old speech-act jargon, 'constatives' rather than 'performatives' (Austin 1962). In other words, his concern is exclusively with whether the judgements match up against their target referents rather than with whatever consequences might flow from the very making of those judgements.

Thus, Tetlock may be guilty of misreading the mass media's preference for pundits who are hedgehogs rather than foxes. These hedgehogs may be lions in disguise, making predictions that are meant to be interventions either to reinforce or to subvert existing political tendencies. Here the media may be less reporting politics than providing an alternative channel for its conduct.

From this standpoint, Tetlock's own disruptive research constitutes a counter-move, one designed precisely to inhibit the hedgehog's more leonine tendencies by luring pundits from their epistemic comfort zones to reveal the limits of their expertise. Of the two species of pundit, Tetlock clearly prefers the fox, not least because the fox's self-consciously limited horizons give it less capacity for causing lasting damage.

Generally speaking, expertise should be seen primarily in sociological rather than strictly epistemological terms (Fuller 1988: chap. 12; Fuller 2002: chap. 3). An expert is no more – and no less – than someone whose word is presumed by other people to decide matters of a certain kind. To be sure, the expert's decision may be underwritten by a coherent body of reliable and relevant knowledge. But such knowledge is neither necessary nor sufficient for expertise to be effective. What matters is that the expert's decision licenses a train of other judgements and actions that attempt to align the world with the decision. Sociologically speaking, expertise is the most potent non-violent form of power available. While there may be resistance to various expert pronouncements, the resistance is channelled in quite specific ways. Experts are clearly marked by their formal training, which opponents must somehow match before issuing a credible challenge, which itself must be conducted by epistemically approved means, typically by mobilizing reasons and evidence in specially designated forums, such as peer-reviewed journals and conferences.

However, it would be a mistake to conclude that 'rule by experts' entails a rigidly governed society. On the contrary, we currently live under such a regime and its conditions are fluid. Given that the expert's power derives from her sphere of discretion, one should expect a variety of judgements and actions to be licensed. For example, two qualified physicians confronted with the same patient may offer quite different diagnoses and treatments. The mark of expert rule is that such variability is tolerated, instead of taken as grounds for charging incompetence on the part of the experts, if not incoherence of the body of expert knowledge itself. This helps explain the socially destabilizing effects of experts challenging each other. Under normal circumstances, experts are presumed competent in their spheres of discretion and are credited for the good consequences of their decisions but, as much as possible, are excused from the bad ones by attributing them to factors legitimately unforeseen by the expert. But this policy is predicated on experts exercising a measure of self-discipline in the kinds of cases they agree to decide. Put bluntly, they need to turn down cases whose initial likelihood of failure is too high.

While the above account of expertise may seem sceptical, a better ancient precedent is provided by the *Stoics*: whereas the sceptics questioned our general ability to define concepts with sufficient certainty to serve as reliable bases for knowledge, the Stoics proposed that our concepts work perfectly well in the paradigm cases for which they were intended but become uncertain beyond that point (Fuller 2005a). The original example that distinguished these two positions remains enshrined in logic textbooks as 'the paradox of the heap', or *sorites* in the original Greek. It is basically the idea that one grain of sand does not constitute a heap, but if you keep adding more grains, at some point the next grain will constitute a heap. Thus, sceptics concluded that 'heap' – and, by extension, all concepts – is hopelessly unclear. For their part, Stoics argued that all that is shown by the paradox is that concepts cannot be applied indefinitely.

Successful experts normally have a Stoic's sense of the conceptual limits within which they should operate. Like Socrates, they know what they do not know. However, much of Tetlock's work is designed to nudge experts outside this epistemic comfort zone, effectively getting them to relax their self-discipline. While Tetlock sees himself as probing the limits of expert knowledge, he is arguably destroying expertise, or at least violating one of its essential preconditions by supposing that an expert in a given field should display a uniform competence in judging all matters logically and counterfactually related to that field – and not simply the range of matters on which the expert would normally be expected to pass judgement.

Tetlock is most explicit on this point when he asks experts to unpack possible futures in terms of finer-grained outcomes, only to discover (perhaps unsurprisingly) that they manage to confer on these specific possibilities greater plausibility than their original judgements of the more generally stated outcomes would logically permit (Tetlock 2005: 189–218). An experiment of this sort reflects Tetlock's implicit understanding of judgment as applied science, that is, the translation of knowledge of potentially universal scope to specific cases. Thus, as Tetlock admits, the hedgehogs are better prepared than the foxes to parry what he calls 'close-call' counterfactual scenarios, in which small perturbations in real conditions would have led to quite different outcomes. Hedgehogs tend to stick to their original predictive frameworks and explain away counterfactuals as 'exceptions that prove the rule' or as products of overactive imaginations or unlikely factors, whereas the foxes are more easily led to admit many more realistic possible outcomes than their theories would seem to allow (212–13). Thus, playing the hedgehog, an international-relations 'realist' (balance-of-power theorist) said, 'I'll change my mind in response to real but not imaginary evidence. Show me an actual case of balancing failing' (212).

The difference seems to be that hedgehogs are firmly anchored in the paradigm cases of their expertise, whereas foxes are always playing with the limiting cases, making them better able to capture the nuances in real-world situations but also more easily disoriented when the divide separating the actual and possible worlds is blurred or erased, as when asked by the experimenter to consider close-call counterfactuals. In a sense, the foxes treat the extra imagination required to specify possible outcomes as if it constituted additional evidence for the likelihood of those outcomes.

Learning to Think the Unthinkable: Post-Truth's Proving Ground

Tetlock's (2003) earlier research into 'thinking the unthinkable' provides a backdrop against which to understand his approach to the limits of expertise. But more important for our purposes, this research pushes experts in a decidedly post-truth direction by forcing them to consider counterfactual scenarios with the same causal and moral seriousness as factual cases. Specifically, when experts are asked to make fine-grained judgements about alternative pasts or futures, they must often entertain, in Tetlock's jargon, *taboo cognitions*: they must trade off a 'sacred' against a 'secular' value. Thus, the sacred value violated in Tetlock's (2005) scenarios is defined by the boundary surrounding an expert's sphere of discretion. The secular experimental setting induces the expert to overstep that boundary, resulting in judgements that effectively humiliate the expert, perhaps even if they turn out to be correct. Some follow-up action – what Tetlock calls 'moral cleansing' – is often needed to re-establish the expert's authority, evidence for which can be found among both foxes and especially hedgehogs, who are inclined to accuse the experimenter of impertinence, if not outright deception, in having got them to relax their normal standards of probative judgement, in particular what is either impossible or inevitable.

The original home for the discussion of moral cleansing is Durkheim's (1964) account of sacred and profane uses of space in society. Not surprisingly, in Tetlock's (2003) earlier thought experiments, devout Christians were asked to place monetary values on the termination of certain human lives or to entertain the psychodynamic consequences for Jesus had he been raised in a single-parent home. Here we need be clear about exactly how Tetlock's thought experiments elicit 'taboo cognitions' from subjects. It is by inviting them to convert a difference in kind to one of degree, or to reduce a qualitative distinction to a quantitative one. When expert historians or pundits are cajoled into exploring possibilities other than those they normally consider permissible, the distinctness of their subject matter starts to disappear – one consequence of which, of course, is that other people start to appear equally expert. For example, once theologians take seriously the idea of pricing human lives, economists are implicitly brought into the conversation. Devote too much concern to Jesus's upbringing and psychoanalysts enter the picture, as in fact they did, courtesy of Albert Schweitzer.

A similar result awaits an ordinary religious believer who ascribes too many human qualities – even their greatest possible versions – to the deity. This is why theological orthodoxy in the Abrahamic faiths has tended to treat God-talk analogically, not literally. Without this semantic boundary work, the biblical claim that humans are created 'in the image and likeness of God' could easily slide into the ceding of theological ground to social scientists, as God comes to be seen as a utopian projection, a superlative human being whose powers we might come to approximate through collective effort. Thus, the 'profane' character of doctrines of human perfectibility and social progress has less to do with their formal opposition to organized religion than their direct competition with it, as in the secular salvation narratives promoted by Comtean positivism and Marxist socialism (Milbank 1990).

In general, we might say that rationality becomes a fully secular mental process once we become adept at making value trade-offs. Decisions that had been previously regarded, in Tetlock's jargon, as 'taboo' (sacred v. profane) or 'tragic' (sacred v. sacred) become simply 'efficient'. Put a bit a more crassly: no end is unconditional – everything is negotiable at a price. Historically this view has been associated with a heightened sense of personal responsibility for the consequences of one's decisions. In other words, reason no longer possesses us – in the sense of compelling a particular outcome to our thinking: rather, we possess reason. An interesting precedent for considering this matter is the social history of the sublimation of violence, which has evolved from our being driven to violence by animal instinct to more calculated inflictions of cruelty in service of some higher goal (cf. Collins 1974).

Explanations – as well as excuses and condemnations – for violent episodes in human history come most easily when they are presented as free-standing 'events' with clear perpetrators and victims, as in the case of 'The Holocaust'. Yet, much of this clarity is an artefact of hindsight created by retrospective accounts that privilege the perspective of one or the other side of the violence. Violence is sublimated when correctives to this historical bias are incorporated into one's own sphere of action. For example, in Nazi Germany, genocide was incrementally instituted over time, the cumulative convergent effect of a policy that was administered by largely indirect means. Awareness of some overall sense of the violence committed was obscured by two features of complex modern societies: the bureaucratization of public administration (whereby each functionary's enjoys discretionary authority within a circumscribed domain, very much like an expert) and language's 'adiaphoric' capacity to refer to something by its defining properties rather than its proper name (Bauman 1995: chap. 6). Together these benchmarks of rational discourse came to be seen as a recipe for systematic dehumanization, once the Nazis proved not to be on the winning side of history. Had the Nazis won, their own rather diffuse – and even 'banal' (à la Eichmann) – understanding of the atrocities they committed would have been normalized, perhaps in the way in which radical poverty in the developing world continues to be (Fuller 2006a: chap. 14).

I make this point, of course, not to rehabilitate the Nazis but to suggest a context in which Tetlock-like exercises in taboo cognition could serve to erode conventional moral intuitions in aid of fostering more advanced moral reflection. So let us think a bit more about the range of politically permissible responses to extreme global poverty, which may well succeed in achieving, in a more diffused and finessed way, what the Nazis arguably aimed for, namely, a very studied form of negligence that manages to subordinate the suffering of potentially identifiable classes of victims to some promised greater good. Here one could manipulate several variables in getting subjects to consider at what point, say, a sweatshop becomes a labour camp, starvation becomes torture and so forth. It is important to stress that the point of such an exercise would not be to induce moral scepticism but to destroy the illusion that 'the immoral' is a realm clearly signposted and hence easily avoided by the scrupulous. Rather, much in the spirit of Existentialism's 'dirty hands' principle, the potential for immorality is present in all judgements, which are ultimately 'arbitrary' in the strict sense of requiring discretion for which one is then personally responsible. The result of such awareness may be to increase our capacity for decisiveness, tolerance and forgiveness.

But even outside the unavoidable controversies surrounding moral judgements, the cultivation of taboo cognitions can have explosive effects. The clearest cases in point may be drawn from the history of science. Imagine a time-travelling Tetlock who asks a sixteenth-century anatomist to think about how the liver might function under conditions that would quite clearly threaten the normal functioning of the human body. For the anatomist, this would be an invitation to countenance a taboo cognition, as he would need to offset the secular value of sheer intellectual curiosity against religiously inspired normative constraints on the practice of anatomy itself, which limited dissections of the human body and, in any case, treated the liver as a proper part of the body, an organ, not a stand-alone piece of organic matter. Of course, the last 500 years have served to secularize the study of the human body so that an anatomist today would find the question quite ordinary. 'Liver' is now defined in more functional than substantive terms – that is, not as something with a particular look, feel, composition or origin, but simply whatever can reliably act as the chemical conversion plant for the human body. In the meanwhile, the human body itself has come to be seen as a self-maintaining system consisting of potentially replaceable parts that are increasingly made to order – and may routinely come to be so, if stem cell research continues to make progress.

I raise this point because the history of anatomy is indicative of what happened to most ancient forms of expertise in the wake of the seventeenth-century Scientific Revolution: their objects lost their sacred boundaries as differences in kind were rendered into ones of degree, such that two states that had been seen as radically different (even violently opposed) – such as movement and rest, living and dead, earthly and stellar, human and animal – came to be seen as two poles of a continuum that may be studied by common means and even experimentally manipulated (Funkenstein 1986). The critical side of the Scientific Revolution conjured up thought experiments very much in the spirit of Tetlock's 'thinking the unthinkable', namely, to elicit contradictory responses at the conceptual edges of existing expertises that served to trigger a fundamental rethinking of – say, in Galileo's case – the nature of motion (Kuhn 1977: 240–65). The net effect of this transformation – the rendering of the constant variable, if you will – was a shift from Aristotle's view of reality as a patchwork of discrete domains of being to Newton's unified vision under which all objects are products of the same set of laws, rendered intelligible by the same set of cognitive processes, albeit operating in different proportions under different conditions (Cassirer 1953).

Here one might consider establishing the experimental discipline of *axioae-tiotics*, whose name is a Greek-rooted neologism for studying the values-causes nexus in people's thought (Fuller 1993: 167–75; cf. Axtell 1993). The premise of this discipline is that even the legitimacy of our basic concepts of epistemic authority – not least those relating to science itself – presuppose a certain understanding of how they came to acquire that authority, the limits of which could be tested if not subverted by presenting historical counterfactuals. Indeed, Kuhn notoriously, but correctly, claimed that practicing scientists needed an 'Orwellian' (aka Whig) understanding of their own history, one that airbrushed all the complexities and alternative trajectories of the past in aid of a streamlined account that justifies the current research frontier being just as it is (Kuhn 1970: 167; cf. Fuller 2000b: introduction). These implicit narratives of legitimation are delicately poised between a hedgehog's overdetermined and a fox's underdetermined view of intellectual history. This normative equilibrium may be easily disturbed, as Tetlock (2005) does in chapters 5–7, by appealing to interpretations of the past that, while not part of the legitimatory lore, nevertheless are taken seriously by professional historians.

I have employed versions of the latter taboo strategy in defending intelligent design theory as an alternative to the neo-Darwinian orthodoxy in biology (e.g. Fuller 2010). In particular, I have stressed that science would not have acquired, and perhaps cannot continue to sustain, its 'view from nowhere' sense of objectivity towards all of reality – and not merely to what is relevant to our species survival – without faith in the idea that we have been specially designed to make sense of it all. This self-understanding of humanity is peculiar to the

Abrahamic religions. In contrast, had Western science been consistently oriented towards species survival, as Darwinists urge, we would not have taken such
a keen interest in, nor conferred so much cultural value on, nor, for that matter,
risked so much on behalf of, the pursuit of reality's physical limits. Darwin's
ablest defender, Thomas Henry Huxley (1893), put the point well when he said
that Newton the Christian had to precede Darwin the apostate – not the other
way round – to explain the zeal with which we have reshaped nature in our own
image, a project that would seem like sheer hubris, if not insanity, had humans
always seen themselves as simply one amongst many species. (Huxley was worried that as Darwinism became part of humanity's self-understanding in the
twentieth century, scientific progress would come to a halt.)

One can imagine presenting Huxley's counterfactual to irreligious scientists today, asking them to imagine how Western science would have reached
the comprehensive achievement of modern physics, had Darwin come to be
accepted first. They would probably think it possible, but it would be interesting to see exactly how they would flesh out the details of an alternate history
that does not depart too far from the main tendencies in the actual history.
In particular, where would one find the motive to conceptualize all of reality
under a finite set of mathematical laws, if one begins with an essentially earth-
bound, species egalitarian view of the natural world? To be sure, a Darwin-first
history would allow for the development of quite sophisticated technologies,
including mathematical techniques, all in aid of human survival across vast
swathes of space and time. Thus, the physical sciences could plausibly reach
the heights of Chinese civilization. But the Chinese did not think it reasonable, or even interesting, to unify all of knowledge under a single intellectual
rubric (Fuller 2015: chap. 6).

Of course, our scientist-subjects might dismiss all this scenario-mongering
as sheer fantasy, as science really stands on its own track record, whatever
theological motives may have been operative in the past. It is a pity, then, that
our sense of science's track record is subject to so much confirmation bias. We
easily recall and even oversell science's empirical and practical successes, while
ignoring or underestimating the costs, failures and outright disasters. Perhaps
the next frontier for 'thinking the unthinkable' is to get experts and lay people
from a variety of backgrounds to draw up balance sheets for science. My guess
is that the resulting track record will look so chequered that science may need
to rekindle its ties to theology to ensure its future legitimacy.

Superforecasting: The Fine Art of Always Preparing for Doomsday

Tetlock has recently introduced the concept of *superforecasting* for forecasters
who seem to have a better-than-average track record at predicting complex

events. His key insight is derived from Moltke, the mastermind behind the newly unified Germany's victory over France in the Franco-Prussian War of 1870–1 and popular icon of Bismarck's Realpolitik (Tetlock and Gardner 2015: chap. 10). For Moltke, effective military strategy operates from a position of neither strength nor weakness but of *resilience*. The strategist must be capable of surviving and improving upon the false predictions that are inevitable in any complex dynamic engagement. Indeed, Moltke famously said that no plan survives first contact with the enemy. Thus improvisation is necessary for the successful realization of any strategy.

In this context, a 'strategy' should be understood as a set of prioritized objectives, the achievement of which can take place by a variety of means, depending on the state of play. This combination of clear overarching aims and discretionary action on the field characterized what Moltke called *Auftragstaktik*, normally translated as 'mission command'. It assumes a sense of utmost trust in both the rightness of the ends set by the commander and the will of the ground troops to achieve them by whatever means it takes. It is an approach that has been facilitated by open (first telegraphic, later telephonic) lines of communication in the period of engagement, conducted in a pyramid structure, whereby the field marshal would process all the different inputs and offer appropriate specific feedback to the ground fighters. The overall process somewhat resembles the internal workings of the brain as proposed in Clark (2016), mentioned earlier in this chapter.

Tetlock's superforecasters embody Moltke's modus operandi in their approach to the future, which is conceptualized as an ontological battlefield open to multiple outcomes, as suggested in the third interpretation of the past's legacy. In terms of counterfactually based philosophical approaches to causation, the successful superforecaster derives from Moltke a sense of the future that is *overdetermined* in the forecaster's favour yet *underdetermined* with regard to obstacles in the way of realizing that future (cf. Fuller 2015: chap. 6). In other words, one should not assume that the opening state of play, including its dominant tendencies, will prevail throughout the conflict. On the contrary, they may be easily reversed, permitting other latent tendencies to reveal themselves. This general frame of mind had led the Prussian army to invent the modern concept of 'war game' (*Kriegsspiel*) with an in-built element of chance. They were basically board game precursors of today's virtual reality simulators which the military has been designing since the 1960s and which are commercially redeployed in the programmes of video games (Mead 2013).

An important lesson taught by the gaming approach to warfare is that the superforecaster must be always open to 'unknown unknowns', to use the decision-theoretic jargon popularized by US Secretary of Defense Donald Rumsfeld during the Iraq war. The key point is that such remote prospects

may be unknown, but they are not *unknowable*. They can be imagined, anticipated and, if not outright defeated, capitalized on. Indeed, one might even
lose the war yet win the peace from the standpoint of overarching objectives.
Such a relatively positive orientation to worst-case scenarios characterizes
the resilient warrior, one always in search of 'room to manoeuvre' (*Spielraum*).
Much depends here on how abstractly one's military objectives are specified.
For example, if the main aim is the long-term prosperity of one's society,
then accepting a defeat that stops short of annihilation followed by a post-
war reconstruction policy funded by the victors may actually make the society
economically stronger within, say, a generation than it was prior to the war.
Arguably that was exactly what happened to Germany and Japan after the
Second World War.

For a deeper understanding of resilience, imagine it as part of a logical
square of opposition, whose founding contraries are the *proactionary* and the
precautionary attitude towards risk, a distinction introduced early in this chapter.
The proactionary embraces risk as an opportunity, the precautionary avoids it
as a threat. The former is associated with the standpoint of the entrepreneur
and adopts a prima facie optimistic perspective on scientific and technological
innovation. The latter is associated with the standpoint of the environmental-
ist and adopts a prima facie pessimistic attitude towards such innovation. Both
proactionaries and precautionaries admit the presence of considerable uncertainty in the world, much – if not most – of it being of human origin, which in
turn implies some distribution of good and bad consequences. However, they
differ in their attitudes to those consequences: The former presumes our world
is *resilient* and the latter *fragile* (cf. Taleb 2012). Thus, the proactionary attitude
towards risk would be refuted by a world in which even a slight change in current conditions would have catastrophic consequences, which in turn would
demonstrate our 'fragility'. In contrast, the precautionary attitude would be
refuted by a world in which even a major change in current conditions would
not only leave us intact but quite possibly stronger, which would demonstrate
our 'resilience'.

What is the value of adopting the 'resilience' perspective? It helps explain
how people have adapted throughout history to complex transformations to
their lifeworld. These have been often accompanied by little fanfare, let alone
understanding or regulation. A case in point in our own time is the centrality of computer-mediated forms of production, consumption and communication, all transpiring within a single generation and epitomized in the
latest smartphone. Nevertheless, specific problems arise as such developments
interact with default expectations of how the world should be. More to the
point: new science and technology tend to cast in a new – and not always flattering or even supporting – light old social, political and economic aspirations,

which can force a major rethink of society's value system. Indeed, we may come to want different things as a result of realizing that we are coming to live in a new world in which our old desires have become obsolete, or otherwise difficult to realize.

A useful concept for exploring people's resilience is what social psychologists call *adaptive preferences*, which is central to the reduction of 'cognitive dissonance', namely, how people maintain a coherent sense of what is meaningful in a world undergoing fundamentally unpredictable changes. After all, even when people fail to predict the future, they still need to see themselves in the future that they failed to predict. An adaptive preference results when aspiration is bent towards expectation in light of experience: we come to want what we think is within our grasp. More than a simple 'reality check', adaptive preference formation involves disciplining one's motivational structure with the benefit of hindsight. Much of what passes for 'wisdom' in life is about the formation of adaptive preferences.

When the social psychologist Leon Festinger suggested the idea in the 1950s, it provided a neat account of how people maintain a sense of autonomy while under attack by events beyond their control (Festinger, Riecken and Schachter 1956). He might have been talking about how the United States and the USSR held their nerve in Cold War vicissitudes, but in fact he was talking about how a millennial religious cult continued to flourish even after its key Doomsday prediction had failed to materialize. A quarter-century later, the social and political philosopher Jon Elster (1983) brilliantly generalized the idea of adaptive preference in terms of the complementary phenomena of 'sour grapes' and 'sweet lemons': we tend to downgrade the value of previously desired outcomes as their realization becomes less likely and upgrade the value of previously undesired outcomes as their realization becomes more likely.

An interesting question is whether adaptive preference formation is rational. Festinger's original case study seemed to imply an answer of no. After a few hours of doubt and despair, the cult regrouped by interpreting the deity's failure to end the world as a sign that the cult had done sufficient good to reverse their fate. This emboldened them to proselytize still more vigorously. One might think that had the cult responded rationally to the failed prophecy, they would have simply abandoned any belief that they were in a special relationship with a higher deity. Instead the cult did something rather subtle. They did not make the obvious 'irrational' move of denying that the prophecy had failed or postponing Doomsday to a later date. Rather, they altered their relationship to the deity, who previously appeared to claim that there was nothing humans could do to reverse their fate. The terms of this renegotiated relationship then gave the cult members a sense of control over their lives which served to renew their missionary zeal.

This is an instance of what Elster called 'sweet lemons', and it is not as obviously irrational as its counterpart, 'sour grapes'. In fact, a 'sweet lemon detector', so to speak, may be a key element of the motivational structure of people who are capable of deep learning from the negative experience of a falsified prediction. Such people come to acquire a clearer sense of what they have truly valued all along so that they are reinvigorated by adversity. The phenomenon of sweet lemons is disorienting to the observer because it highlights just how much we presume that others share our overarching values. We do not simply respect the autonomy of others. We also expect, somewhat paradoxically, that by virtue of their autonomy they will become more like us over time. Thus, the post-prophecy behaviour of Festinger's cult is confusing because they carried on in a version of what they had previously done. In psychological terms, they projected a second-order 'ideal' sense of self that could then be dissociated from the first-order 'empirical' self whose actions have been found wanting. Put plainly, they learned from experience, but what they learned was to become more like themselves.

Adaptive preferences are arguably scalable, perhaps to become the properties of entire cultures and species. Indeed, humanity may be improving its capacity for adaptive preference formation, a sign of our increasing resilience. It is striking that the disruption and destruction resulting from faulty forecasts do little to prevent similar precipitating behaviours in the future. However, we seem to be getting better each time at converting liabilities into virtues, making it easier 'to lose the war and win the peace'. In this respect, adaptive preference formation serves much the same purpose as the 'relative advantage of backwardness' has served in economic history – about which more below. Both phenomena are informed by the idea – in the one case, self-induced and in the other, delivered by history – that falling a step behind may be the best strategy to get two steps ahead.

Here the sociologist Max Weber's *theodicy* of suffering offers some insight (Joas 2000: chap. 2). Theodicy is the theological discipline that attempts to explain God's curious sense of justice, which seems to permit considerable suffering even among the blameless. Weber distinguishes two ways in which the world religions sublimate the pain suffered from thwarted expectations that good behaviour will be rewarded. One, rather like Elster's sour grapes, is ressentiment, whereby pain results in the devaluation of a prior aspiration; the other, rather like his sweet lemons, is *ecstasy*, which operates in reverse, investing value in a state that one had previously tried to avoid. In the latter case, a glimpse of the truth is revealed in the moment of pain, a light at the end of the tunnel, which offers an incentive to carry on with greater resolve. 'Perseverance' was the name given to this attitude in the early modern period, which applied to, say, the hardships suffered by the Puritan founders of the

United States. But since that time it has been extended to cover the entire scientific world view, courtesy of Popper (1963). This helps explain the curious Popperian logic whereby the relentless falsification of scientific hypotheses does not serve to falsify scientific enquiry itself (i.e. scepticism) but, on the contrary, demonstrates that science is getting closer to the truth.

Both sides of Weber's theodicy of suffering are at play in secular guise in the history of technological innovation. Take the case of automotive transport. Within a half-century of mass-produced cars, the hazards that were forecast at the time of their invention had been realized: cars were a major – if not the major – source of air and noise pollution. The roads required for cars ravaged the environment and alienated their drivers from nature. Yet none of that seemed to matter – or at least not enough to lead people to abandon automobiles. Rather, car production worldwide has continued apace while becoming a bit more environment-friendly to avoid the worst envisaged outcomes. For better or worse, the value package that Henry Ford and others were selling in the early twentieth century still holds: we value the car's freedom and speed not only over the connectedness to nature provided by the horse but also the relatively low ecological impact offered by mechanized public transport today. Sour grapes and sweet lemons combine to raise the car to a liberator of the human spirit from 'nature', now seen as the imposer of constraints. Whether this turns out to be a good long-term survival strategy remains an open question.

We have seen that forecasting carries emotional investments in both the process and the outcome. The appeal of Tetlock's superforecasting is that one might improve predictive capacity by adopting a productive attitude towards error, perhaps even welcoming it as an incentive to raise one's game and divest oneself of unnecessary commitments. The agents in both Moltke's military strategy and Festinger's millenarian cult enact this policy on the field and in the mind by a clear division between steadfast ends and fluid means, the 'ends justifies the means', as the Jesuits originally put it. Put another way, they *self-instrumentalize*, which is to say, they prioritize the future they want to such an extent that along the way they are willing to trade off features of their present selves that are not essential to this future's realization. In metaphysical terms, the trail left in the wake of this process identifies the past with the discarded 'matter' and the future with the emerging 'mind'. Or, in the terms often used to discuss the change in world view involved in the Scientific Revolution, the unified 'substance' of the self is reduced to a set of 'functions' (Cassirer 1953).

At a more psychological level, superforecasting may be understood as a form of *cognitive immunology*. In other words, when one forecasts with an eye to one's own fate over a range of possible futures (not only the one predicted as most likely), then it becomes easier to change tack as the future is unfolding so that one's ultimate goal is realized. The rethink required involves a reversal of what economists have recognized as the 'hyperbolic' time-discounting tendencies of consumers, whereby they fixate on short-term goals even when these compromise their long-term ones (Ainslie 1992). One especially creative – albeit daunting – way of understanding this psychological reversal was put forward by the Cold War RAND Corporation strategist Herman Kahn, who coined the phrase 'thinking the unthinkable' (1962).

Kahn's signature phrase built on his earlier work on scenarios involving the aftermath of a thermonuclear war in which, say, 25–50 per cent of the world's population is wiped out over a relatively short period of time (1960). How do we rebuild humanity under those circumstances? This is not so different from 'the worst-case scenarios' proposed nowadays under conditions of severe global warming. Kahn's point was that we need now to come up with the relevant new technologies that would be necessary the day after doomsday. Moreover, he believed that such a strategy was likely to be politically more tractable than trying actively to prevent doomsday through, say, unilateral nuclear disarmament, which would simply make us more vulnerable to attack from those who failed to disarm.

Kahn's understanding of the politics of the situation had predictive value as a self-fulfilling prophecy. The United States largely followed Kahn's advice. And precisely because doomsday never happened, we ended up in peacetime with the riches that we have come to associate with Silicon Valley, a major beneficiary of the US federal largesse during the Cold War (Mazzucato 2013). The Internet was developed as a distributed communication network in case the more centralized telephone system were taken down during a nuclear attack. In this respect, Stanley Kubrick was quite prescient in subtitling his 1964 black comedy film *Dr. Strangelove*, whose eponymous character was partly modelled on Kahn's mindset, *How I Learned to Stop Worrying and Love the Bomb*.

Truth be told, Kahn's 'ahead of the curve' thinking has been characteristic of military-based innovation generally, for which in recent times the Defense Advanced Research Projects Agency (DARPA) has served as both torchbearer and flack catcher (Belfiore 2009). As we have seen in the references to Moltke throughout this book, warfare focuses minds on what is dispensable and what is necessary to preserve – and indeed, how to enhance that which is necessary to preserve. Here one can truly say that 'necessity is the mother of invention'. We win even – and especially – if doomsday never happens.

Doomsday scenarios invariably invite discussions of human resilience and adaptability. A good starting point is a distinction drawn in cognitive archaeology between *reliable* and *maintainable* artefacts (Bleed 1986). Reliable artefacts tend to be 'overdesigned', which is to say, they can handle all the anticipated forms of stress, but most of those never happen. Maintainable artefacts tend to be 'underdesigned', which means that they make it easy for the user to make replacements when disasters strike, which are assumed to be unpredictable. While in principle, resilience and adaptability could be identified with either position, the Cold War's proactionary approach to doomsday implies the latter (Fuller and Lipinska 2014: 35–36). In other words, we want a society that is not so dependent on the most likely scenarios – including the most likely negative ones – that we could not cope in the event of a very unlikely, very negative scenario. The knowability of Rumsfeld's 'unknown unknowns' is clearly a guiding assumption here.

Tetlock's taboo cognition research offers a good basis for thinking about such extreme scenarios to make superforecasting a normal feature of our mode of being in the world. Candidates for the relevant 'unknown unknowns' in today's world come from the interaction effects of relatively independent scientific and societal trends, each individually benign yet capable of producing malign emergent consequences. These would constitute the relevant 'taboos'. Let us say that techno-optimists are vindicated and that global warming does not turn out to be as disruptive to current socio-economic trends as techno-pessimists think. Yet these same optimists also tend to predict an increase in both the labour force's technological redundancy and human life extension. In that case, an extreme scenario might be to imagine both trends in tandem. Social scientists following Tetlock's lead could then ask both expert and lay subjects to pinpoint the negativity in such a scenario, including trade-offs that might redeem the negative aspects. In this way, the value of taboo topics ranging from 'universal basic income' to more liberal policies concerning suicide and euthanasia would be likely given voice – and we could start to prepare to live in a world where worst-case scenarios are no longer seen as a threat to our 'normal' way of life.

Conclusion: From Superforecasting to Precipitatory Governance

Even in a post-truth world, someone needs to take responsibility for what happens. By 'responsible', I mean decisive in the outcome of an uncertain situation: the one who took the decision that made the most difference. However, the term's meaning tends to be spun towards the moral standpoint of an observer after the fact. Thus, agents may be deemed 'irresponsible' if they did not decide matters in the 'right' way. The phrase 'responsible innovation', popular in EU circles, is proposed in just this spirit. It captures the idea of being wise before the fact, when 'the fact' consists in suboptimal, if not catastrophic, impacts for a broad range of constituencies in the wake of some proposed innovation. In that case, one tries to anticipate those consequences with an eye to mitigating if not avoiding them altogether; hence, a policy of *anticipatory governance* (Barben et al. 2008).

Anticipatory governance is the territory of the precautionary principle, according to which innovations with great capacity for harm – regardless of benefits – would not be introduced at all (Schomberg 2006). The EU itself implements a more moderate version which recognizes the generally beneficial character of innovation, while insisting on monitoring its effects on society and the larger environment. Implied here is that one might have one's cake and eat it: innovations would be collectively owned to the extent that those potentially on the receiving end would be encouraged from the outset to voice their concerns and even opposition, which will shape the innovation's subsequent development (Schomberg 2013).

But recalling Kahn, one needs to be responsible not only before the fact but also after the fact, especially when 'the fact' involves suboptimal impacts, including 'worst-case scenarios'. This is the opposite of anticipatory governance. Call it *precipitatory governance*. Precipitatory governance operates on the assumption that some harm will be done, no matter what course of action is taken, and that the task is to derive the most good from it. I say 'derive the most good' because I do not wish to limit the range of considerations to the mitigation of harm or even to the compensation for harm, though I have dealt with those matters elsewhere (Fuller and Lipinska 2014: chap. 4). In addition, as we have seen under the rubric of 'thinking the unthinkable', the prospect of major harm may itself provide an opportunity to develop innovations that would otherwise be regarded as unnecessary – if not outright utopian – to the continuation of life as it has been.

Precipitatory governance proposes to extend Kahn's doomsday turn of mind into peacetime. After all, the most disruptive innovations in peacetime – such as the automobile or the personal computer – tend to be seen as 'necessary' only after the fact, once they have systematically reconfigured the market. Before the fact, they are often seen as speculative, if not risky. Indeed, only in the second half of the nineteenth century, once industrial capitalism had colonized the popular imagination, did the very word 'innovation' begin to acquire unequivocally positive connotations. Prior to that time, 'innovation' was more often than not seen as a synonym for 'monstrosity' (Godin 2015). In any case, capitalism has increasingly enabled entrepreneurs to make headway with their innovations through the exploitation of vulnerabilities in the market leaders. Historically a major source of those vulnerabilities has been the market leaders' relative lack of resilience to changes in aspects of the environment over which they have little control. This is the spirit in which to get into precipitatory governance.

Suppose the current climate science consensus is correct that the earth's average temperature will rise by at least two degrees centigrade by the end of this century. Which human groupings, institutions, technologies and so forth are likely to survive and possibly flourish under these conditions, and which ones not? Those placed in the latter category should be seen as ripe for entrepreneurial investment now – not necessarily to prevent the global temperature rise but to ensure that the most overall benefit is gained from a situation that is both highly likely and prima facie suboptimal. The source of that benefit would be the removal of some of the preconditions that had made those now endangered people and practices so adaptive in the past. In effect, the impending catastrophe will have broken what sociologists call 'traditions' and economists call 'path dependencies', which had made a 'business as usual' attitude so attractive for so long.

An interesting economic precedent for this general line of thought is what the mid-twentieth century Harvard economic historian Alexander Gerschenkron (1962) called 'the relative advantage of backwardness'. His basic idea was that, in principle, each successive nation could industrialize more quickly by learning from its predecessors without having to follow in their footsteps. The 'learning' amounts to innovating more efficient means of achieving and often surpassing the predecessors' level of development. The 'advantage' here comes from not having to bear the burden of a particular past, such as the well-documented path to industrialization taken by England, starting with the Enclosure Acts in the seventeenth century. Put another way, a newcomer nation can see more clearly the 'superstitious' elements of its predecessors' self-understanding of their success, as the newcomer is forced to sort out the 'wheat' from the 'chaff' of that narrative for purposes of designing its own innovation strategy.

A post-catastrophic humanity would be similarly positioned to benefit from Gerschenkron's sense of 'backwardness' on a global scale vis-à-vis the pre-catastrophic humanity – provided that the record of human knowledge remains relatively intact in the wake of even the greatest catastrophe. To be sure, this is not a trivial assumption. But ongoing projects of digital archiving and curation of everything from artefacts to organisms make it an increasingly plausible one, regardless of one's views on humanity's exposure to *existential risk*, which is nowadays the fashionable face of doomsday.

The idea of 'existential risk' capitalizes on the prospect of a very unlikely event that, were it to pass, would be extremely catastrophic for the human condition. Thus, the high value of the outcome psychologically counterbalances its low probability. It is a bit like Pascal's Wager, whereby the potentially negative consequences of not believing in God – to wit, eternal damnation – rationally compels you to believe in God, despite your instinctive doubts about the deity's existence. The Oxford transhumanist philosopher Nick Bostrom (2015) has popularized 'existential risk' in this sense with regard to computational power outpacing our ability to control it. However, this line of reasoning underestimates both the weakness and the strength of human intelligence. On the one hand, we are not so powerful as to create a 'weapon of mass destruction', however defined, that could annihilate all of humanity; on the other, we are not so weak as to be unable to recover from whatever errors of design or judgement that might be committed in the normal advance of science and technology in the human lifeworld. In this context, precipitatory governance is an antidote to the species defeatism exemplified by the promotion of existential risk.

Whenever military history is written with the wisdom of hindsight, precipitatory governance is often implied. Thus, those who like their winners to remain winners and losers to remain losers find it disconcerting to learn, say, how much West Germany and Japan benefitted from having been on the losing end of the Second World War, such that by the 1960s they were global socio-economic trendsetters. Nevertheless, once time has healed enough wounds, such a turn of events is generally seen as having been a good thing, even if the conditions precipitating them were not. Moreover, the increasing economic rationalization of warfare, starting with the Second World War, has shifted the horizon of precipitatory governance from hindsight to foresight. Indeed, our times have witnessed the rise of private and corporate investors on the lookout to capitalize on major catastrophes, natural or human in origin. An exemplar of this tendency is the Halliburton Company, which specializes in infrastructure projects in oil-rich regions of the world, places in periodic need of reconstruction due to their susceptibility to military conflict. One can decry Halliburton's sweetheart deals and adventurism, yet admit that in their self-serving way they are in the business of precipitatory governance.

Behind this relatively sanguine attitude towards the potential benefits of catastrophe is a controversial idea, namely, that the competences associated with creation and destruction are fundamentally similar, or in Joseph Stalin's homely quip, 'You can't make an omelette without breaking some eggs'. The 1940s served to sharpen this point, not only in Schumpeter's (1942) popularization of the concept of 'creative destruction' as the key to entrepreneurial innovation, the lifeblood of capitalism's endless rejuvenation by reconfiguring new markets from old ones. (A quarter-century earlier Schumpeter had introduced the idea in the context of Ford's transport revolution.) But also, and more potent at the time, J. Robert Oppenheimer's recalled the Hindu deity Shiva, the 'creator and destroyer of worlds', upon the detonation of the first atomic bomb in what remains the most apt description of nuclear energy ever given.

In both creation and destruction, there are two key elements in the application of force: concentration of resources and strategic focus. Of course, destruction is potentially 'creative' only to those who are empowered in its wake. Or, as economists would put it, the exploitation of any opportunity entails opportunity costs, very much in the spirit of the third interpretation of the past's legacy earlier in the chapter: to create one world is to delay, if not outright destroy, the prospects for another world. But 'creative destruction' may equally suggest an ability or willingness to see a deeper benefit in the apparent harm done to these alternative futures, perhaps by expanding one's horizons to enable one's ideals to be realized even in a world that was not of one's choosing. The ethical and political challenge facing the advocate of precipitatory governance is to make credible this intuition, which amounts to recasting the experience of permanent material harm as a failure of imagination. I have discussed this challenge in terms of *moral entrepreneurship*, a kind of applied theodicy (Fuller 2011: chap. 5; Fuller 2012: chap. 4). It is admittedly a tough sell.

Nevertheless, it far from fanciful. Indeed, this proposed conversion of harm to good is bound to become increasingly relevant, as the sorts of disruptions that in the past would have required military aggression or entrepreneurial genius can now be licensed – if not outright induced – on the basis of the sort of projective research that is routinely used in forecasting. For example, MIT's Computer Science and Artificial Intelligence Laboratory has recently estimated that all of New York City's taxis could be replaced by 3,000 carpooling vehicles of the sort that are already in use by Uber, thereby improving traffic flows and environmental quality (Cooke 2017). The downside of such a proposal is that it would effectively eliminate the livelihood of taxi drivers. To be sure, many things can be done in response to such research, not least trying to prevent its ever becoming policy. This may involve commissioning some countervailing research, which would be in the spirit of turning science into the currency of warfare. However, the disruptive policy may simply include plans for redeploying the displaced taxi drivers, showing that it would be to society's greater benefit for them not to do what they were originally trained to do. This would be to treat science more as an agent of divine justice – a kind of applied theodicy – in which the drivers would suddenly come to lose the comparative advantage that they previously enjoyed through no fault of their own. Clearly then, the fairness of this policy will hang on what passes for 'redeployment'.

At this point, it is worth observing that precipitatory governance constitutes 'responsible innovation' in quite a deep yet elusive sense. It is natural to hear in the phrase 'responsible innovation' the idea that innovation might be irresponsibly undertaken, which of course is something to be avoided. However,

I hear something different, namely, that innovation is inevitable and that the challenge is to extend the range of 'responsibility' throughout the entire innovation process. More to the point: insofar as innovation itself is seen as central to the promotion of the human condition, then it would be irresponsible *not* to provide for the promotion of innovation even in light of cases where particular innovations have led to disruptive, perhaps even catastrophic, outcomes. In effect, precipitatory governance is an insurance policy against any anti-innovation backlash in the wake of a major disruption: it fulfils our responsibility not only *for* but also *to* innovation.

Precipitatory governance is ultimately underwritten by the *proactionary principle*, the opposite of the precautionary principle, introduced at the start of this chapter. If the precautionary principle commands 'Do no harm!', the proactionary principle commands 'No pain, no gain!' My ultimate guide here is Jean-Paul Sartre, who argued that we all have 'dirty hands' regardless of whether our actions are seen as having caused a net harm or a net benefit to the world. Even the most precautionary policies that aim to lower the level of risk in the environment – say, by prohibiting genetically modified organisms or lowering carbon emissions – incur opportunity costs with regard to the development of science and technology, which also need to be put on humanity's balance sheet when assessing the prospects of future generations. After all, the harm 'prevented' by having taken certain courses of action (e.g. to lower environmental risk) is no less speculative than the harm 'caused' by not having taken certain courses of action (e.g. to promote science and technology). Both rely on counterfactuals. Nevertheless, we tend to accord greater realism to the former sort of counterfactual, perhaps due to its tie-in to action explicitly taken in this world. Generally speaking, such intuitive asymmetry has favoured the precautionary principle. However, the law is sensitive to the cognitive bias involved here, which it tries to redress by devoting considerable attention to culpability through negligence.

To be sure, precipitatory governance appeals to a broader conception of negligence than is normally entertained in the law, which is indicative of a post-truth approach to normativity more generally. In particular, its concerns extend to the failure to promote the innovative capacity of human beings. This idea is not entirely without precedent in the law. Consider the original justifications for patents and copyrights in the early modern period. They presumed that complacency is the ultimate enemy of prosperity, which can only be met by the explicit encouragement of enterprise (Fuller 2002: chap. 2). As we saw in Chapter 4, this sensibility is historically tied to the biblical idea that humanity exists in a fallen state that requires some sort of remediation. Modernity has been largely about enterprise and innovation overtaking faith and piety as the primary modes of remediation. Thus, we witness a shift in the modal standing of innovation: from something barely permitted to something incentivized if not outright obligated. It amounts to regarding the 'status quo' or 'received wisdom' not as a fount of authority and stability but as always under suspicion for frustrating the efforts of people to live up to potential.

Let us conclude with an example. Consider policies that aim to provide 'equal opportunities' for each successive generation. The United Negro College Fund famously promoted this idea in its 1971 US ad campaign as 'A mind is a terrible thing to waste'. A generation of blacks who might be otherwise left behind needs to be given a 'head start' by providing them with skills for a future whose fundamental uncertainty is interpreted positively as a level playing field, in which default forms of advantage (and disadvantage) need no longer be in effect. This is, of course, a more recognizably 'progressive' route to reach largely the same state of mind into which Kahn frightened cold warriors, in which the nuclear 'ground zero' creates the level playing field. Looming in the background of both scenarios is the moral harm caused by the failure to cultivate human potential, which is taken to be one of great negligence, the negative consequences of which accrue to both the neglecter and the neglected, given that default forms of advantage and disadvantage need not persist in an ever-changing world. Thus, in the United Negro College Fund campaign, the failure of blacks to integrate into American society becomes the failure of whites to integrate them. Ultimately it is a failure of the imagination of the whites to envisage how much better things might be, if the short term strategy of 'good enough' – what at the start of the chapter was called 'satisficing' – did not serve so easily as a proxy for whatever overarching good we might be striving for.

THE ARGUMENT IN A NUTSHELL

The post-truth condition is not simply a product of our times but endemic to the history of Western thought, as originally expressed in the Platonic Dialogues. Moreover, post-truth is not a condition limited to politics but extends to science as well. Indeed, the post-truth condition enables us to see more clearly the complementarity of politics and science as spheres of thought and action. Each in its own way is involved in a struggle for 'modal power', namely, control over what is possible.

This book explores the field of play common to politics and science through such concepts as the 'military-industrial will to knowledge', 'Protscience' and 'precipitatory governance'. What makes these concepts 'post-truth' is their refusal to take the rules of the epistemic game at face value. There is more to knowledge than the consensus of expert opinion, and even what the experts take as knowledge need not be interpreted as the experts would wish.

While some have decried recent post-truth campaigns that resulted in victory for Brexit and Trump as 'anti-intellectual' populism, they are better seen as the growth pains of a maturing democratic intelligence, to which the experts will need to adjust over time. The lines of intellectual descent that have characterized disciplinary knowledge formation in the academy might come to be seen as the last stand of a political economy based on rent-seeking, whereby credentials are seen more as arresting than facilitating the circulation of knowledge in society.

One need not pronounce on the specific fates of, say, Brexit or Trump to see that the post-truth condition is here to stay. The post-truth disrespect for established authority is ultimately offset by its conceptual openness to previously ignored people and their ideas. They are encouraged to come to the fore and prove themselves on this expanded field of play. In this respect the post-truth condition marks a triumph of democracy over elitism, albeit one that potentially tilts the balance towards 'chaos' over 'order'.

Plato had realized that the only way to curb democracy's chaotic tendencies was to ensure that discussions about the rules of the game and one's position in the game are kept separate – what logicians have respectively called 'first-order' and 'second-order' questions. We can debate one or the other but not both at once. That is the 'truth condition'. The post-truth condition challenges that assumption, as players jockey for position in the current game, while at the same they try to change the rules so as to maximize their own overall advantage.

In a post-truth utopia, both truth and falsehood are themselves democratized. Thus, no one will ever be deemed incorrigible – either incorrigibly right or incorrigibly wrong. While this prospect seems quite egalitarian, before signing the contract you should attend to the fine print. You will neither be allowed to rest on your laurels nor rest in peace. You will always be forced to have another chance to play in a game whose rules are forever contestable. Perhaps this is why some people still prefer the truth condition, no matter who sets the rules of the game.

GLOSSARY

Adaptive Preference – The tendency for people's desires to bend towards their expectations in light of experience, as a strategy of reducing 'cognitive dissonance'. On the one hand, this phenomenon may simply reveal people's failure to confront their mistakes squarely; on the other, it may show that people learn over time what really matters to them, which helps them go forward. The truth condition takes the former position, the post-truth condition the latter.

Anticipatory Governance – A growth industry within the field of science and technology studies (STS) that organizes focus groups, wiki media and other platforms to gauge people's attitudes to innovations that are predicted to happen in the short- to medium-term future. While the STS researchers themselves are normally content to stress the ambivalence if not fear that people now exhibit to the future, their paymasters operate in the spirit of market research to adjust the content and branding of the forthcoming innovation to ensure maximum public uptake.

Anti-expert Revolution – This is the more general phenomenon to which the coinage of 'post-truth' responds. It is marked by the concurrent rise in greater access to education and information, on the one hand, and deeper suspicion of political and scientific authority, on the other.

Antirealism –The official philosophy of what in this book is called the 'post-truth condition'.

Aristotle (384–322 BCE) – Student of Plato who was the most influential philosopher in the West until the seventeenth century. Instead of banning the playwrights as purveyors of alternative versions of reality, as Plato wished, he called for the plot lines of plays to be resolved by their end, thereby clearly sealing them off as 'fiction', as opposed to the 'fact' that exists outside of the theatre.

Bacon, Francis (1561–1626) – The personal lawyer to King James I of England, who provided the first explicit account of the scientific method in terms of not only its internal workings but also its potential for societal transformation. An admirer of continental Europe's inquisitorial style of

adjudication, he saw nature as a hostile witness playing a game of conceal-
ment against humanity, which in turn reflected our biblically 'fallen' state.

Bernays, Edward (1891–1995) – The great nemesis of Walter Lippmann,
with whom he served in Woodrow Wilson's successful public relations campaign to get Americans to fight in the First World War. Bernays favoured
a democratized and capitalized approach to public relations, which he
turned into a billion-dollar industry based on market research fuelled by
an experimental approach to impression management. Many of the techniques inspired by his work, including focus groups, passed into mainstream
social science.

Brexit – Britain's departure from membership in the European Union (EU),
which was decided in a UK-wide referendum in 2016. The referendum
provided the largest voter turnout in British history, but the outcome was
seen by political and scientific elites as surprising and deleterious to both
the United Kingdom and the EU. Yet, it is also reasonably seen as a profound democratic rebuff of expert authority.

Chronos – The sense of time required of any successful 'lion'. It presents
history as an orderly succession of events, which cumulatively legitimizes
the status quo, and in terms of which any external disruption is perceived
as an existential risk.

Consensus, Scientific – An ideologically inflated understanding of the fact
that the largely self-organizing and self-certifying group of people called
'scientists' decide collective ownership of pieces of knowledge that some of
their number have produced based on an academic 'peer-review' process,
which is then presented to the rest of society as humanity's best guess at
the truth.

Creative Destruction – The phrase coined by the economist Joseph
Schumpeter for the revolutionary impact of innovation on the market.
A genuine innovation does not simply add another competitor to an already
existing field of play. It creates a new field of play, thereby forcing the players to recast themselves or disappear. Its dynamic very much resembles that
of a Kuhnian 'paradigm shift', except that 'entrepreneurs' (Schumpeter's
name for the innovators) are more deliberate in creatively destroying a target market than so-called 'scientific revolutionaries', who usually only enjoy
the title in retrospect.

Customized Science – A discriminating attitude to science that is characteristic of the post-truth condition, which is made possible by both
increased education and more widespread access to information sources.
Closely related to Protscience, its epistemology is marked by a clear distinction between what one 'knows' (that is, has learned) and what one 'believes'
(that is, acts upon).

Double Truth Doctrine – Plato's original idea that social order is maintained by the promulgation of two truths, one for the elites who determine the rules of the game and one for the masses who obey the rules. These are the 'post-truth' and 'truth' conditions, respectively, as discussed in this book.

Existential Risk – The fashionable face of doomsday in the post–Cold War era. Instead of nuclear holocaust, one envisages either global climate catastrophe or the onslaught of superintelligent machines. Whereas those who promote the idea of existential risk normally think of it as something to avoid under all circumstances, there is a post-truth interpretation, somewhat in the spirit of Herman Kahn, to regard it as an opportunity for 'out of the box' thinking in innovation policy.

Expertise – The sort of knowledge that a 'scientific consensus' allegedly bestows, what Kuhn called 'normal science'. On this basis, universities can extract the form of rent known as 'tuition', in return for which 'credentials' are dispensed, somewhat in the spirit of Catholic indulgences – except that the university diploma conferring expertise absolves one only of prior ignorance of a given field. Protscience regards expertise as a protection racket.

Fake News – A calling card of the post-truth condition, whereby the contesting parties accuse each other of imposing the wrong conceptual framework for telling what is true and false. While the phrase is normally associated with Donald Trump, its spirit is traceable at least to the polemics of the Protestant Reformation.

Foxes (à la Erasmus) – Philip Tetlock's term for pundits whose predictive accuracy is restricted to a succession of short-term forecasts and hence fail to see larger emerging tendencies. It is based on Erasmus's distinction between the 'fox', who knows many little things, and the 'hedgehog', who knows one big thing. Put in Machiavellian terms: because foxes lack the default legitimacy of the lions, they must be always proving themselves, which limits their capacity for seeing deeply into the future.

Foxes (à la Machiavelli) – Vilfredo Pareto's term for those elites who believe that they can increase their advantage by relaxing Plato's double truth doctrine, thereby allowing more people to potentially determine the rules by which truth itself is determined. In conventional political terms, foxiness is the stance of the 'opposition', whose calling card is 'change' and 'new blood'.

Galilei, Galileo (1564–1642) – The original poster boy for the post-truth condition, at least according to philosopher Paul Feyerabend. Although Galileo is now an iconic figure of the Scientific Revolution, at the time he strategically overstated his knowledge claims, gambling that others would find the claims sufficiently compelling to create the conditions under which they might be confirmed. And he was proved right.

Game, Science as a – The post-truth understanding of science, whereby it is more important to know the conditions under which a knowledge claim can be true or false than to know whether it is true or false. In that sense, science is a game for the control of what in this book is called 'modal power'.

Haber, Fritz (1868–1934) – The consummate interdisciplinary post-truth scientist whose chemical innovations, deployed in radically different fields of play, have arguably had the most impact on more people's lives in all of history – both in terms of the numbers who have been allowed to live and the numbers who have been made to die.

Hedgehogs – Philip Tetlock's term for pundits who stick to predicting the same thing under all circumstances, which in the long term gives them a spurious air of consistency, which serves to amplify the significance of the few times they turn out to be right.

Hollywood Knowledge – A double-edged phrase that can mean either the 'hyperrealism' of motion pictures, which leads people to confuse on-screen fiction with off-screen fact, or the hybrid form of knowledge produced by scientists and other academics in consultation with motion picture studios in developing a film, which in turn may serve to fuel public and even professional perceptions of what is scientifically possible. The post-truth condition promotes the latter and regards the former as its vehicle.

Information Overload – A concept invoked periodically since the late Middle Ages to capture not only the proliferation of texts but also the inability of readers to decide how to act in relation to them. In the early modern period, this led to more personalized approaches to knowledge, as epitomized in the work of Jean-Jacques Rousseau. However, in recent years, a more polarized approach has emerged, according to which most of what passes for information is really either noise or content we have yet to fully exploit (aka 'undiscovered public knowledge').

Intelligent Design Theory – A scientifically updated version of creationism that is a paradigmatic Protscience and part of the anti-expert revolution that characterizes the post-truth condition.

Interdisciplinarity – A form of enquiry that ideally both borrows from and sees beyond different academic disciplines. More often than not, its impetus has come from outside the academy, especially what this book calls the 'military-industrial will to knowledge'.

Intersectionality – The recognition that people have multiple identities, some of which benefit or harm them more than others. In the modern era, states have maintained social order by stabilizing such identity markers as race, class and gender. However, the post-truth era is about the devolution of this capacity to self-organizing 'identity politics', resulting in a vision of society as a plurality of cross-cutting power games, in which the identity of the players is potentially as fluid as the identity of the games that they are playing.

Kahn, Herman (1922–1983) – Leading RAND Corporation Cold War analyst who advocated 'thinking the unthinkable', by which he meant making the most of a 'limited nuclear war' by starting now to develop innovations that would suit a world in which fundamental aspects of our current infrastructure were destroyed. The point is that even if 'doomsday' does not happen, the innovations would still be there. The Internet was a product of this line of thought. Stanley Kubrick's classic 1964 film, *Dr. Strangelove*, was partly inspired by Kahn.

Kairos – The sense of time ('timing') required of any successful 'fox'. It is the moment when applying the available force will have the maximum effect, thereby changing the field of play decisively, ideally to one's advantage. In its aftermath, all of the prior history is rewritten to legitimize the new, revolutionary order.

Kant, Immanuel (1724–1804) – Generally regarded as the cornerstone figure in modern Western philosophy, his great achievement was to formally recognize that even if God's existence cannot be proven, nevertheless there is a God-like hole in our minds that is filled by second-order thinking, which is the source of what in this book is called 'modal power'.

Kuhn, Thomas (1922–1996) – The most influential theorist of science in the second half of the twentieth century, whose cyclical account of science as alternating between a long 'normal' and a short 'revolutionary' phase corresponds to what in this book are called, respectively, the 'truth' and 'post-truth' conditions, the former dominated by 'lions' and the latter by 'foxes'.

Lions – Vilfredo Pareto's term, adapted from Niccolò Machiavelli, for those elites who aim to increase their advantage by adhering strictly to Plato's double truth doctrine, which favours those willing and able to play by the rules by which truth is currently determined. In conventional political terms, the 'establishment' strikes the leonine pose, upholding the virtues of 'stability' and 'tradition'.

Lippmann, Walter (1889–1974) – Probably the twentieth century's best exemplar of Plato in practice, having helped craft America's dominant self-image on the world stage from Woodrow Wilson to Richard Nixon. He spoke early against Edward Bernays's efforts to marketize public relations as a machine to 'manufacture consent'. He wanted a more centralized and circumspect approach to informing the public, which contributed to his reputation as the godfather of 'professional' or 'objective' journalism.

Military-Industrial Will to Knowledge – A counterweight to the default disciplinary norms of academic inquiry. As the name 'military-industrial' suggests, this form of enquiry is more explicitly organized and goal oriented, typically vis-à-vis improving some aspect of the human condition. Its knowledge characteristically aims to be 'productive', implying at once efficiency and efficacy. Private foundations were the exemplars of this modus operandi in the twentieth century.

Modal Power – Control over what can be true or false, which is reflected in intuitions about what is possible, impossible, necessary and contingent. In the history of philosophy, this form of knowledge has been called 'transcendental', though its source has been much disputed. For Plato, it involved the imposition of the will of the philosopher-king. In the Christian era, it has required a journey into the mind of the divine creator, with science playing an increasingly important role in the modern era. However, Kant famously argued that our sense of the transcendental simply reflects the limits of our own minds, which in turn has served to fuel the post-truth condition.

Moltke, Baron Helmuth von (1800–1891) – One of the great military commanders of all times, he masterminded the newly unified Germany's defeat of France in the Franco-Prussian War. His doctrine of 'permanent emergency', whereby in peacetime a nation should be preparing to fight the next war, is the inspiration for 'superforecasting' and 'precipitatory governance'.

Pareto, Vilfredo (1848–1923) – A founding figure of neoclassical political economy and sociology, whose influence until the 1960s arguably rivalled that of Karl Marx. Pareto updated Machiavelli for the world of modern liberal democracy, not least through his theory of the 'circulation of elites' ('lions' versus 'foxes'), which captures the dynamic of the post-truth condition.

Pareto's 80/20 Principle – Eighty per cent of the observed effects are the result of 20 per cent of the available causes. Pareto claimed that this principle prevailed in so many aspects of so many societies that it amounted to a natural law. In particular, no matter the society's source of wealth production, 80 per cent of it will be in the hands of 20 per cent of the people. The principle seems to apply to academic knowledge production as well, and the question is whether it constitutes simply a brute hierarchy (the 'lion' view) or an exploitable resource (the 'fox' view).

Pasteur's Quadrant – Named for the modus operandi of nineteenth-century biochemist Louis Pasteur, who pursued various enquiries aimed at discovering the fundamental principles needed to improve human life in areas of most concern to France. In the future, such 'use-oriented basic research' might come to be understood in the context of genetics, as eugenics tried to do in the twentieth century. In any case, it subverts the academic myth that basic research must precede any useful applications. The military-industrial will to knowledge flourishes in Pasteur's Quadrant.

Peer Review – A piece of academic Newspeak to refer to the relatively few academics who pass judgement on most academics' work, on the basis of which career advancement and public credibility are determined. Yet it operates as a desultorily organized protection racket, which is surprising given the trust that both academics and non-academics invest in the process.

Plato (427–347 BCE) – The canonical founding figure of Western philosophy, who already recognized the need for aspiring rulers to adopt a post-truth attitude to reality, according to which authority is exerted through the exercise of modal power. His corpus consists largely of simplified dramas, or 'dialogues', in which Socrates figures as the protagonist. In some of those dialogues, Socrates argues that the ideal polity should censor dramatic performance, a seminal moment in the history of the double truth doctrine.

Popper, Karl (1902–1996) – The natural heir to John Stuart Mill, perhaps the only twentieth-century philosopher whose contributions to both political and scientific thought were equally great, centring on the idea of the 'open society'. Although not fully appreciated at the time, Popper played the 'fox' in calling for 'permanent revolution' in science when responding to Thomas Kuhn's more 'leonine' account of scientific change in the 1960s.

Post-Truth Condition – A social order whose members are always and everywhere thinking both in terms of what game to play and what move to make in whatever game might be in play. Such a society suspends the default assumption that its members share a common reality, and hence a common sense of truth conditions. Plato believed that a well-ordered society had to restrict the post-truth condition to its ruling class, who then decides the terms of a common reality that apply to everyone else.

Precautionary Principle – A risk-averse approach to rationality which prioritizes avoiding harm over doing good. Enshrined in much European health and environmental legislation, the principle presumes that high risk corresponds to high threat and hence aims for a 'sustainable' world in which successive generations are provided with comparable resources and opportunities to those of their predecessors.

Precipitatory Governance – A political strategy inspired by Herman Kahn, which regards 'worst-case scenarios' (aka 'existential risks') not as things to be avoided at all cost but as opportunities, the exploitation of which may yield benefits even if the imagined scenarios never come to pass.

Proactionary Principle – A risk-seeking approach to rationality which prioritizes doing good over avoiding harm. Its presumption that high risk corresponds to high opportunity has informed both the entrepreneurial ethos and the sort of 'revolutionary' science championed by Karl Popper. The rationale is that while a glowing success is preferred, outright failure potentially teaches more – to the survivors – than muddling through.

Protestant Reformation – A formalization of doctrinal differences in Western Christianity through the proliferation of clerical authorities that begins in the sixteenth century, which in turn provides the theological backdrop for the rise of secular modernity. It also introduced some of the signature moves of the post-truth condition, which can be seen at play in the emergence of Protscience.

Protscience – A contraction of 'Protestant science', it covers disparate anti-establishment science movements that have been often cast as 'pseudoscience', including creationism and New Age medicine. Their followers share a desire to integrate science more directly into their own lives, very much in the spirit of Christianity's Protestant reformers. The result is what in this book is called 'customized science'.

Psephology – The social scientific study of voter behaviour, normally understood as a special case of public opinion research, which is in turn indebted to market research of the sort championed by Edward Bernays. As the subjects of psephology have become more aware of the social scientists studying them, the subjects have also become adept in their responses, increasingly confounding the psephologists, as in the cases of Brexit and Trump's election.

Realism – The official philosophy of what in this book is called the 'truth condition'.

Relative Advantage of Backwardness – A phrase coined by the economic historian Alexander Gerschenkron to describe the advantage of late industrial nations that can learn from their predecessors without having to repeat their history. This can result in innovations that expedite development so that the latecomers end up leapfrogging over their predecessors. Precipitatory governance sees any major catastrophe as offering just such an opportunity for those who survive the catastrophe.

Rent-Seeking – The feature of feudalism that is anathema to both capitalists and socialists, it consists in the capacity to generate value by making a resource – be it land or labour – less rather than more productive. This condition is made possible by the monopoly nature of ownership, which in turn allows 'rent' to be charged prior to any use of the resource. As long as potential users have no alternative but to go through the owner to get what they want, rents can be charged. In the post-truth condition, expertise – and perhaps academic knowledge more generally – is regarded as a form of rent-seeking.

Revisionism – The rewriting of history to one's advantage, which reflects the strategic function of controlling the sort of hold that the past is allowed to exert on the future. It happens even in fields, such as philosophy, mathematics and much of the natural sciences, which present themselves as being ahistorical in nature. In the post-truth condition, foxes are keenest to revise history, but the lions are keenest to keep a particular revision in place.

Rousseau, Jean-Jacques (1712–1778) – The French Enlightenment thinker who has survived most intact as a 'philosopher' in today's world. He holds the enviable title of being both democracy's best advocate and critic. For him, democracy was possible only if everyone – including the rulers – agreed on the rules of the game. He believed that this required common values, instantiated in a 'general will'. Rousseau saw self- and landownership very much in this light, which has made him a darling of both 'communists' and 'communitarians'. However, he offered a post-truth escape route in his innovation of personal expression in literature.

Saint-Simon, Count Henri de (1760–1825) – Paradigmatic 'utopian socialist' in contrast to Karl Marx and Friedrich Engels's own 'scientific socialism', he was the most consistent opponent to rent-seeking before Marx. He is arguably the originator of 'knowledge management' and the 'knowledge society', stressing a sense of efficiency and productivity normally lacking in academia. Inspired by Francis Bacon, he was perhaps the first to think self-consciously in terms of the military-industrial will to knowledge.

Science and Technology Studies (or STS) – An interdisciplinary offshoot of the sociology of knowledge, whose pioneers were mainly intellectual carpetbaggers. In its original social constructivist phase during the final quarter of the twentieth century, it was widely seen as 'postmodernist' in orientation and 'demystifying' of science. We would now say that it was a harbinger of the post-truth condition. However, in recent years it has tried, with only partial success, to row back from that initial position.

Second-Order Thought – What is popularly called 'going meta', including 'metalanguage', 'metalogic' and 'metapsychology'. It involves thinking about the rules of the game one is playing, especially assuming that several rules are possible, in terms of which a given play may be valid. Second-order thought is the default state of mind of someone in the post-truth condition.

Social Constructivism – A twentieth-century intellectual movement across a wide range of disciplines that stresses an 'antirealist', and hence 'post-truth', approach to knowledge production. While social constructivism is naturally seen as involving a 'sociological' approach to knowledge, that was only made explicit in the 1960s with the publication of Peter Berger and Thomas Luckmann's *The Social Construction of Reality*.

Socrates (d. 399 BCE) – Perhaps the most iconic figure in Western philosophy, he is known mainly in his role as Plato's avatar in most of the Platonic Dialogues. Socrates's defence of 'truth' against his Sophist opponents amounts to outfoxing them on their own post-truth turf.

Sophists – Foreign traders in classical Athens whose distinctive wares – 'rhetoric' as Plato dubbed them – captivated the natives. Socrates appears in Plato's Dialogues as someone who could match the Sophists' salesmanship, but who argued the true value of their wares lay in their 'proper' use. Thus, Socrates outfoxed them in the sort of second-order thought that constitutes the post-truth condition.

Spielraum – 'Room for manoeuvre' in nineteenth-century German military jargon. It inspired Max Weber's idea of *Verstehen*, whereby a constrained space provides the opportunity for one to exercise a distinctive form of freedom that may serve to expand – or contract – their freedom in the future. The space of possibilities in these situations can be studied objectively by social scientists.

Superforecasting – Political psychologist Philip Tetlock's term, inspired by Baron Helmuth von Moltke, for gaming the future to one's advantage by assuming that one might suffer significant short-term losses in pursuit of the ultimate objective. Thus, the stress is placed not on getting each forecast correct but on extracting the most out of any forecast if it turns out to be false. It offers a productive spin on adaptive preferences and underwrites precipitatory governance and the proactionary principle.

Taboo Cognition – Political psychologist Philip Tetlock's term for the sort of thinking involved in situations where subjects must suspend one of their most cherished ('sacred') assumptions in order to address a hypothetical problem. Such situations may be seen as experimentally inducing the sort of 'thinking the unthinkable' mentality that Herman Kahn advocated more generally during the Cold War.

Theodicy – The justification of 'evil' (in its broadest sense to include natural catastrophes) in the world, given the undisputed existence of a good God. A hotbed of rationalist theology in early modern Europe, theodicy frequently reappears in secular guise, especially in narratives concerning the 'logic' of capitalism and Darwinian evolution. Its main relevance to the post-truth condition lies in the prospect for what I call 'moral entrepreneurship', the fine art of converting liabilities into virtues – that originally Godlike skill which is also central to precipitatory governance.

Truth Condition – The idea of reality as a single agreed playing field, with science understood as its signature game. In this book, it is the opposite of the post-truth condition.

Undiscovered Public Knowledge – A phrase coined by the library scientist Don Swanson in the 1980s for the vast bulk of academic publications that are underutilized – even by academics – if not ignored altogether, notwithstanding their untapped epistemic value, which nevertheless has been exploited by military and industrial concerns throughout the modern era.

Vaihinger, Hans (1852–1933) – The founder of modern Kant scholarship who in the early twentieth century advanced a philosophy of 'fictionalism', which is a precursor of contemporary post-truth thinking. His basic idea was that the actual world is the world towards which we act 'as if' it were real.

Veritism – An epistemological stance in contemporary analytic philosophy that is so keen to disconnect knowledge of the truth from any sense of human agency that a veritist could possess knowledge by reliably uttering truths, even without any understanding of how she or he manages to achieve this feat. The most that can be said on its behalf is that it clearly inhibits the sort of second-order thought that informs the post-truth condition.

Weber, Max (1864–1920) – A founder of sociology as an academic discipline, who understood the complementarity of political and scientific rationality in broadly post-truth terms.

Wikipedia – The online encyclopaedia, appearing in multiple languages, which is undoubtedly the largest collective intellectual project ever successfully undertaken – both in terms of people involved and topics tackled. Notwithstanding its considerable epistemic shortcomings, *Wikipedia* has the strongest claim to being an instance of universal democratic knowledge production.

Wittgenstein, Ludwig (1889–1951) – Influential twentieth-century analytic philosopher whose career conveniently divides into two halves, corresponding to what in this book is called the 'truth' (*Tractatus Logico-Philosophicus*) and the 'post-truth' (*Philosophical Investigations*) conditions.

REFERENCES

Ainslie, G. (1992). *Picoeconomics: The Strategic Interaction of Motivational States within the Individual.* Cambridge, UK: Cambridge University Press.

Anderson, C. (2006). *The Long Tail.* New York: Hyperion.

Aron, R. (1967). *Main Currents in Sociological Thought II.* Garden City, NY: Anchor Doubleday.

Artaud, A. (1958). *The Theatre and Its Double.* New York: Grove Weidenfeld.

Austin, J. L. (1962). *How to Do Things with Words.* Cambridge, MA: Harvard University Press.

Axtell, G. (1993). 'In the Tracks of the Historicist Movement: Re-Assessing the Carnap-Kuhn Connection'. *Studies in the History and Philosophy of Science* 24: 119–46.

Baker, E. and Oreskes, N. (2017). 'Science as a Game, Marketplace or Both: A Reply to Steve Fuller'. *Social Epistemology Review and Reply Collective* 6(9): 65–69.

Barben, D., Fisher, E., Selin, C. and Guston, D. (2008). 'Anticipatory Governance of Nanotechnology: Foresight, Engagement and Integration'. In *Handbook of Science and Technology Studies*, edited by E. Hackett, O. Amsterdamska, M. Lynch and J. Wajcman, 979–1000. Cambridge, MA: MIT Press.

Barber, B, ed. (1970). *L. J. Henderson on the Social System.* Chicago: University of Chicago Press.

Bauman, Z. (1995). *Life in Fragments: Essays in Postmodern Morality.* Oxford: Blackwell.

Belfiore, M. (2009). *The Department of Mad Scientists.* New York: HarperCollins.

Bell, A. (2010). 'Taking Science Journalism "Upstream"'. (3 September). http://alicerosebell .wordpress.com/2010/09/03/taking-science-journalism-upstream/. Accessed 25 March 2018.

Bell, D. (1960). *The End of Ideology.* New York: Free Press.

Berger, P. and Luckmann, T. (1966). *The Social Construction of Reality.* Garden City NY: Doubleday.

Berle, A. and Means, G. (1932). *The Modern Corporation and Private Property.* New York: Harcourt Brace and World.

Bernal, J. D. (1939). *The Social Function of Science.* London: Macmillan.

Bernays, E. (1923). *Crystallizing Public Opinion.* New York: Boni Liveright.

——— . (1928). *Propaganda.* New York: Horace Liveright.

Bernstorff, J. v. and Dunlap, T. (2010). *The Public International Law of Hans Kelsen.* Cambridge, UK: Cambridge University Press.

Bew, J. (2016). *Realpolitik: A History.* Oxford: Oxford University Press.

Birch, K. (2017). 'Financing Technoscience: Finance, Assetization and Rentiership'. In *The Routledge Handbook of the Political Economy of Science*, edited by D. Tyfield, R. Lave, S. Randalls and C. Thorpe, 169–81. London: Routledge.

Blair, A. (2010). *Too Much to Know: Managing Scholarly Information before the Modern Age.* New Haven, CT: Yale University Press.

Bleed, P. (1986). 'The Optimal Design of Hunting Weapons: Maintainability or Reliability?' *American Antiquity* 51: 737–47.

Bloom, H. (1973). *The Anxiety of Influence*. Oxford: Oxford University Press.

Bloor, D. (1976). *Knowledge and Social Imagery*. London: Routledge & Kegan Paul.

Bostrom, N. (2015). *Superintelligence*. Oxford: Oxford University Press.

Brynjolfsson, E. and McAfee, A. (2014). *The Second Machine Age*. New York: W. W. Norton.

Burawoy, M. (2005). 'For Public Sociology'. *American Sociological Review* 70: 4–28.

Cadwalladr, C. (2017). 'The Great British Brexit Robbery: How Our Democracy Was Hijacked'. *The Guardian*. (7 May). https://www.theguardian.com/technology/2017/may/07/the-great-british-brexit-robbery-hijacked-democracy. Accessed 25 March 2018.

Cassirer, E. ([1910] 1953). *Substance and Function*. New York: Dover.

Chandler, A. (1962). *Strategy and Structure*. Cambridge, MA: MIT Press.

Charles, D. (2005). *Master Mind: The Rise and Fall of Fritz Haber*. New York: Ecco.

Chopra, D. (1989). *Quantum Healing: Exploring the Frontiers of Mind/Body Medicine*. New York: Bantam.

Chopra, D. and Tanzi, R. (2012). *Super Brain: Unleashing the Explosive Power of Your Mind*. New York: Harmony Books.

Church, G. and Regis, E. (2012). *Regenesis: How Synthetic Biology Will Reinvent Nature and Ourselves*. New York: Basic Books.

Clark, A. (2016). *Surfing Uncertainty: Prediction, Action and the Embodied Mind*. Oxford: Oxford University Press.

Collins, H. (1985). *Changing Order: Replication and Induction in Scientific Practice*. London: Sage.

Collins, H. and Evans, R. (2007). *Rethinking Expertise*. Chicago: University of Chicago Press.

Collins, R. (1974). 'Three Faces of Cruelty: Towards a Comparative Sociology of Violence'. *Theory, Culture and Society* 1: 415–40.

———. (1979). *The Credential Society*. New York: Academic Press.

———. (1995). 'Prediction in Macrosociology: The Case of the Soviet Collapse'. *American Journal of Sociology* 100 (6): 1552–93.

———. (1998). *The Sociology of Philosophies: A Global Theory of Intellectual Change*. Cambridge, MA: Harvard University Press.

———. (2004). *Interaction Ritual Chains*. Princeton: Princeton University Press.

Cooke, L. (2017). 'MIT Says 3000 Ride-Sharing Cars Could Replace All New York City Taxis'. Inhabitat.com (Posted 3 January). http://inhabitat.com/mit-says-3000-ride-sharing-cars-could-replace-all-new-york-city-taxis/. Accessed 25 March 2018.

Crenshaw, K. (1991). 'Mapping the Margins: Intersectionality, Identity Politics, and Violence against Women of Color'. *Stanford Law Review* 43: 1241–99.

Csiszar, A. (2017). 'From the Bureaucratic Virtuoso to Scientific Misconduct: Robert K. Merton, Eugene Garfield, and Goal Displacement in Science'. Paper delivered to annual meeting of the History of Science Society (Toronto: 9–12 November).

Darnton, R. (1984). *The Great Cat Massacre and Other Episodes in French Cultural History*. New York: Basic Books.

Dennett, D. (1995). *Darwin's Dangerous Idea*. New York: Simon and Schuster.

Deutsch, K. (1963). *The Nerves of Government*. New York: Free Press.

Duff, A. (2012). *A Normative Theory of the Information Society*. London: Routledge.

Dummett, M. (1978). *Truth and Other Enigmas*. London: Duckworth.

Dupré, J. (1993). *The Disorder of Things*. Cambridge, MA: Harvard University Press.

Durkheim, E. (1964). *Elementary Forms of the Religious Life*. (Orig. 1912) New York: Collier Macmillan.

Eisenstein, E. (1979). *The Printing Press as an Agent of Change*. 2 vols. Cambridge, UK: Cambridge University Press.

Elster, J. (1978). *Logic and Society*. Chichester, UK: John Wiley & Sons.

———. (1979). *Ulysses and the Sirens*. Cambridge, UK: Cambridge University Press.

———. (1983). *Sour Grapes: Studies in the Subversion of Rationality*. Cambridge, UK: Cambridge University Press.

———. (2000). *Ulysses Unbound*. Cambridge, UK: Cambridge University Press.

Eppler, M. and Mengis, J. (2004). 'The Concept of Information Overload'. *Information Society* 20: 325–44.

Etzkowitz, H. (2002). *MIT and the Rise of Entrepreneurial Science*. London: Routledge.

Evans, D. (2003). *Placebo: Mind over Matter in Modern Medicine*. London: Harpercollins.

Festinger, L., Riecken, H. and Schachter, S. (1956). *When Prophecy Fails*. Minneapolis: University of Minnesota Press.

Feuer, L. (1974). *Einstein and the Generations of Science*. New York: Basic Books.

Feyerabend, P. (1975). *Against Method*. London: Verso.

———. (1981). *Realism, Rationalism and the Scientific Method*. Cambridge, UK: Cambridge University Press.

Feynman, R. (1974). 'Cargo Cult Science'. *Engineering and Science* 37(7): 10–13.

Firth, N. (2009). 'The Politics Factor: Simon Cowell Unveils Plan to Launch Election Debate Show'. *Daily Mail* (London). 15 December.

Fricker, M. (2007). *Epistemic Injustice*. Oxford: Oxford University Press.

Fuller, S. (1988). *Social Epistemology*. Bloomington: Indiana University Press.

———. (1993). *Philosophy of Science and Its Discontents*. 2nd ed. (Orig. 1989). New York: Guilford Press.

———. (1996). 'Recent Work in Social Epistemology'. *American Philosophical Quarterly* 33: 149–66.

———. (1997). *Science*. Milton Keynes, UK: Open University Press.

———. (2000a). *The Governance of Science*. Milton Keynes, UK: Open University Press.

———. (2000b). *Thomas Kuhn: A Philosophical History for Our Times*. Chicago: University of Chicago Press.

———. (2002). *Knowledge Management Foundations*. Woburn, MA: Butterworth-Heinemann.

———. (2003a). *Kuhn vs Popper: The Struggle for the Soul of Science*. Cambridge, UK: Icon.

———. (2003b). 'In Search of Vehicles for Knowledge Governance: On the Need for Institutions That Creatively Destroy Social Capital'. In *The Governance of Knowledge*, edited by N. Stehr, 41–76. New Brunswick, NJ: Transaction Books.

———. (2005a). 'Philosophy Taken Seriously But Without Self-Loathing'. *Philosophy and Rhetoric* 38(1): 72–81.

———. (2005b). *The Intellectual*. Cambridge, UK: Icon.

———. (2006a). *The New Sociological Imagination*. London: Sage.

———. (2006b). *The Philosophy of Science and Technology Studies*. London: Routledge.

———. (2007a). *Science vs Religion?* Cambridge, UK: Polity.

———. (2007b). *New Frontiers in Science and Technology Studies*. Cambridge, UK: Polity.

———. (2008). *Dissent over Descent*. Cambridge, UK: Icon.

———. (2009). *The Sociology of Intellectual Life: The Career of the Mind in and around the Academy*. London: Sage.

———. (2010). *Science: The Art of Living*. Durham, UK: Acumen.

———— . (2011). *Humanity 2.0: What It Means to Be Human Past, Present and Future*. London: Palgrave Macmillan.

———— . (2012). *Preparing for Life in Humanity 2.0*. London: Palgrave Macmillan.

———— . (2013). 'Deviant Interdisciplinarity as Philosophical Practice: Prolegomena to Deep Intellectual History'. *Synthese* 190: 1899–916.

———— . (2015). *Knowledge: The Philosophical Quest in History*. London: Routledge.

———— . (2016a). *The Academic Caesar: University Leadership Is Hard*. London: Sage.

———— . (2016b). 'Morphological Freedom and the Question of Responsibility and Representation in Transhumanism'. *Confero* 4(2): 33–45.

———— . (2017). 'The Social Construction of Knowledge' In *The Routledge Companion to Philosophy of Social Science*, edited by L. McIntyre and A. Rosenberg, 351–61. London: Routledge.

Fuller, S. and Collier, J. (2004). *Philosophy, Rhetoric and the End of Knowledge*. 2nd ed. (Orig. 1993 by Fuller). Hillsdale, NJ: Lawrence Erlbaum.

Fuller, S. and Lipinska, V. (2014). *The Proactionary Imperative: A Foundation for Transhumanism*. London: Palgrave.

———— . (2016). 'Is Transhumanism Gendered? The Road from Haraway'. In *The Future of Social Epistemology*, edited by J. Collier, 237–46. Lanham, MD: Rowman & Littlefield.

Funkenstein, A. (1986). *Theology and the Scientific Imagination*. Princeton: Princeton University Press.

Gallie, W. B. (1956). 'Essentially Contested Concepts', *Proceedings of the Aristotelian Society* 56: 167–98.

Gehlen, A. (1988). *Man: His Nature and Place in the World*. (Orig. 1940). New York: Columbia University Press.

Gerschenkron, A. (1962). *Economic Backwardness in Historical Perspective*. Cambridge, MA: Harvard University Press.

Gibbons, M., Limoges, C., Nowotny, H., Schwartzman, Scott P. and Trow, M. (1994). *The New Production of Knowledge*. London: Sage.

Gilbert, W. (1991). 'Towards a Paradigm Shift in Biology'. *Nature*, 10 January, 349(6305): 99.

Godin, B. (2015). *Innovation Contested: The Idea of Innovation over the Centuries*. London: Routledge.

Goldacre, B. (2012). *Bad Pharma*. London: Harpercollins.

Goldman, A. (1999). *Knowing in a Social World*. Oxford: Oxford University Press.

Goodman, N. (1955). *Fact, Fiction and Forecast*. Cambridge, MA: Harvard University Press.

———— . (1978). *Ways of Worldmaking*. Indianapolis: Hackett.

Green, D. A. (2017). '"In Some Possible Branches of the Future Leaving Will Be an Error" – An Exchange about Brexit with Dominic Cummings'. *Jack of Kent Blog* (4 July). http://jackofkent.com/2017/07/in-some-possible-branches-of-the-future-leaving-will-be-an-error-an-exchange-about-brexit-with-dominic-cummings/. Accessed 25 March 2018.

Gregory, J. and S. Miller (2000). *Science in Public: Communication, Culture and Credibility*. London: Perseus

Haack, S. (1978). *Philosophies of Logics*. Cambridge, UK: Cambridge University Press.

Habermas, J. (1971). *Knowledge and Human Interests*. (Orig. 1968). Boston: Beacon Press.

Halevy, E. (1928). *The Growth of Philosophic Radicalism*. London: Faber and Faber.

Hamblin, C. L. (1970). *Fallacies*. London: Methuen.

Hammond, K. and Stewart, T., eds. (2001). *The Essential Brunswik*. Oxford: Oxford University Press.

Hayek, F. (1945). 'The Use of Knowledge in Society'. *American Economic Review* 35: 519–30.

Heims, S. J. (1991). *Constructing a Social Science for Postwar America.* Cambridge, MA: MIT Press.

Henderson, M. (2012). *The Geek Manifesto.* London: Bantam.

Herf, J. (1984). *Reactionary Modernism.* Cambridge, UK: Cambridge University Press.

Herman, E. and Chomsky, N. (1988). *Manufacturing Consent: The Political Economy of the Mass Media.* New York: Random House

Hershberg, J. (1993). *James B. Conant: Harvard to Hiroshima and the Making of the Nuclear Age.* Palo Alto, CA: Stanford University Press.

Horgan, J. (1996). *The End of Science.* Reading, MA: Addison-Wesley.

Hughes, R. (1993). *The Culture of Complaint.* Oxford: Oxford University Press.

Humphreys, P. (2004). *Extending Ourselves: Computational Science, Empiricism and the Scientific Method.* Oxford: Oxford University Press.

Huxley, T. H. (1893). 'Evolution and Ethics.' Romanes Lecture. Oxford (18 May) http://aleph0.clarku.edu/huxley/CE9/E-E.html.

Isaac, J. (2011). *Working Knowledge: Making the Human Sciences from Parsons to Kuhn.* Cambridge, MA: Harvard University Press.

Jackson, B. (2009). 'At the Origins of Neo-Liberalism: The Free Economy and the Strong State, 1930–47', *Historical Journal* 53: 129–51.

Jansen, S. C. (2013). 'Semantic Tyranny: How Edward L. Bernays Stole Walter Lippmann's Mojo and Got Away with It and Why It Still Matters'. *International Journal of Communication* 7: 1094–111.

Jasanoff, S. (1990). *The Fifth Branch: Science Advisors as Policy Makers.* Cambridge, MA: Harvard University Press.

Jaszi, P. (1994). 'On the Author Effect: Contemporary Copyright and Collective Creativity'. In *The Construction of Authorship: Textual Appropriation in Law and Literature*, edited by M. Woodmansee and P. Jaszi, 29–56. Durham, NC: Duke University Press.

Jeffries, S. (2011). 'Brian Cox: "Physics Is Better than Rock'n'Roll"'. *The Guardian* (London). 24 March.

Joas, H. (2000). *The Genesis of Values.* Chicago: University of Chicago Press.

Johnson, A. (2017). 'Why Brexit Is Best for Britain: The Left-Wing Case'. *New York Times* (28 March). https://www.nytimes.com/2017/03/28/opinion/why-brexit-is-best-for-britain-the-left-wing-case.html.

Johnson, C. (2012). *The Information Diet: The Case for Conscious Consumption.* Sebastopol, CA: O'Reilly Media.

Johnson, P. (1997). *Defeating Darwinism by Opening Minds.* Downers Grove, IL: Varsity Press.

Kahn, H. (1960). *On Thermonuclear War.* Princeton: Princeton University Press.

———. (1962). *Thinking about the Unthinkable.* New York: Horizon Press.

Kaldor, M. (1982). *The Baroque Arsenal.* London: Deutsch.

Kane, P. (2016). 'Leading Brexiteers Are Pining for All of Us to Embrace a Life Built on Havoc'. *The National* (Glasgow: 18 June). http://www.thenational.scot/comment/14867243.Pat_Kane__Leading_Brexiteers_are_pining_for_all_of_us_to_embrace_a_life_built_on_havoc/. Accessed 25 March 2018.

Kass, L. (1997). 'The Wisdom of Repugnance'. *New Republic,* June 2.

Kauffman, S. (2008). *Reinventing the Sacred: A New View of Science, Reason and Religion.* New York: Basic Books.

Kay, L. (1993). *The Molecular Vision of Life: Caltech, the Rockefeller Foundation and the Rise of the New Biology.* Oxford: Oxford University Press.

Kevles, D. (1992). 'Foundations, Universities and Trends in Support for the Physical and Biological Sciences, 1900–1992'. *Daedalus* 121(4): 195–235.

Khanna, P. (2017). 'To Beat Populism, Blend Democracy and Technocracy, Singapore Style'. *Straits Times*, 21 January. http://www.straitstimes.com/opinion/to-beat-populism-blend-democracy-and-technocracy-spore-style. Accessed 25 March 2018.

Kirby, D. (2011) *Lab Coats in Hollywood: Scientists Impact on Cinema, Cinema's Impact on Science and Technology*. Cambridge, MA: MIT Press.

Knight, D. (2006). *Public Understanding of Science*. London: Routledge.

Kohler, R. (1994). *Partners in Science: Foundations and Natural Scientists*. Chicago: University of Chicago Press.

Kuhn, T. (1970). *The Structure of Scientific Revolutions*. 2d ed. (Orig. 1962). Chicago: University of Chicago Press.

——— . (1977). *The Essential Tension*. Chicago: University of Chicago Press.

Langley, P., Simon, H., Bradshaw, G. and Zytkow, J. (1987). *Scientific Discovery*. Cambridge, MA: MIT Press.

Latour, B. (1987). *Science in Action*. Milton Keynes, UK: Open University Press.

——— . (1993). *We Have Never Been Modern*. Cambridge, MA: Harvard University Press.

——— . (2004). 'Why Has Critique Run Out of Steam? From Matters of Fact to Matters of Concern'. *Critical Inquiry* 30(2): 225–48.

Lecourt, D. (1976). *Proletarian Science?* London: Verso.

Lepenies, W. (1988). *Between Literature and Science: The Rise of Sociology*. Cambridge, UK: Cambridge University Press.

Lessig, L. (2001). *The Future of Ideas: The Fate of the Commons in a Connected World*. New York: Random House.

Lippmann, W. (1922). *Public Opinion*. New York: Harcourt, Brace and Company.

——— . (1925). *The Phantom Public*. New York: Macmillan.

Lynch, W. (2001). *Solomon's Child*. Palo Alto, CA: Stanford University Press.

Maslow, A. (1998). *Maslow on Management*. New York: John Wiley & Sons.

Mayer, J. (2016). *Dark Money: The Hidden History of the Billionaires behind the Rise of the Radical Right*. London: Penguin.

Mazlish, B. (1989). *A New Science: The Breakdown of Connections and the Birth of Sociology*. Oxford: Oxford University Press.

Mazzucato, M. (2013). *The Entrepreneurial State*. London: Anthem Press.

McCloskey, D. (1982). *The Rhetoric of Economics*. Madison: University of Wisconsin Press.

——— . (2006). *The Bourgeois Virtues: Ethics for an Age of Commerce*. Chicago: University of Chicago Press.

——— . (2010). *Bourgeois Dignity: Why Economics Can't Explain the Modern World*. Chicago: University of Chicago Press.

McKenzie, R. and Tullock, G. (2012). *The New World of Economics*. 6th ed. (Orig. 1975). Berlin: Springer.

McLuhan, M. (1951). *The Mechanical Bride*. New York: Vanguard Press.

Mead, C. (2013). *War Play: Video Games and the Future of Conflict*. Boston: Houghton Mifflin Harcourt.

Melzer, A. (2014). *Philosophy between the Lines: The Lost History of Esoteric Writing*. Chicago, IL: University of Chicago Press.

Merton, R. (1965). *On the Shoulders of Giants*. New York: Free Press.

——— . (1968a). *Social Theory and Social Structure*. 3rd ed. (Orig. 1949) New York: Free Press.

——— . (1968b). 'The Matthew Effect in Science'. *Science* 159(3810): 56–63.

——— . (1976). *Sociological Ambivalence and Other Essays*. New York: Free Press.

Milbank, J. (1990). *Theology and Social Theory*. Oxford: Blackwell.

Mirowski, P. (2002). *Machine Dreams: How Economics Became a Cyborg Science*. Cambridge, UK: Cambridge University Press.

Morozov, E. (2013). *To Save Everything, Click Here*. London: Allen Lane.

Neumann, M. (2006). 'A Formal Bridge between Epistemic Cultures: Objective Possibility in the Times of the Second German Empire'. In *Foundations of the Formal Sciences: History of the Concept of the Formal Sciences*, edited by B. Löwe, V. Peckhaus and T. Räsch, 169–82. London: Kings College Publications.

Nida, E. (1964). *Towards a Science of Translation*. The Hague: E. J. Brill.

Nisbett, R. and Ross, L. (1980). *Human Inference: Strategies and Shortcomings of Human Judgement*. Englewood Cliffs, NJ: Prentice-Hall.

Oreskes N. and Conway E. M. (2011). *Merchants of Doubt: How a Handful of Scientists Obscured the Truth on Issues from Tobacco Smoke to Global Warming*. New York: Bloomsbury.

Parsons, T. (1937). *The Structure of Social Action*. New York: McGraw-Hill.

Penrose, R. (1989). *The Emperor's New Mind*. Oxford: Oxford University Press.

Peoples, L. (2010). 'The Citation of Wikipedia in Judicial Decisions'. *Yale Journal of Law and Technology* 12(1): 1–51.

Pettegree, A. (2005). *Reformation and the Culture of Persuasion*. Cambridge, UK: Cambridge University Press.

Phillips, A. (2017). 'Playing the Game in the Post-Truth Era'. *Social Epistemology Review and Reply Collective* (30 June). https://social-epistemology.com/2017/06/30/playing-the-game-in-a-post-truth-era-amanda-phillips/. Accessed 25 March 2018.

Pickering, A. (2010). *The Cybernetic Brain*. Chicago: University of Chicago Press.

Polanyi, K. (1944). *The Great Transformation*. New York: Farrar and Rinehart.

Popper, K. (1957). *The Poverty of Historicism*. New York: Harper and Row.

———. (1963). *Conjectures and Refutations*. London: Routledge & Kegan Paul.

———. (1981). 'The Rationality of Scientific Revolutions'. In *Scientific Revolutions*, edited by I. Hacking, 80–106. Oxford: Oxford University Press.

Porter, T. (1986). *The Rise of Statistical Thinking, 1820–1900*. Princeton: Princeton University Press.

Postman, N. (1985). *Amusing Ourselves to Death: Public Discourse in the Age of Show Business*. New York: Penguin.

Poulakos, J. (1990). 'Interpreting Sophistical Rhetoric: A Response to Schiappa'. *Philosophy and Rhetoric* 23: 218–28.

Prendergast, C. (1986). "Alfred Schutz and the Austrian School of Economics." *American Journal of Sociology* 92: 1–26.

Price, D. (1963). *Little Science, Big Science*. New York: Columbia University Press.

Prigogine, I. and Stengers, I. (1984). *Order out of Chaos*. New York: Bantam.

Proctor, R. (1991). *Value-Free Science? Purity and Power in Modern Knowledge*. Cambridge, MA: Harvard University Press.

Putnam, H. (1978). *Meaning and the Moral Sciences*. London: Routledge.

Rabinbach, A. (1990). *The Human Motor: Energy, Fatigue and the Origins of Modernity*. New York: Basic Books.

Rawls, J. (1971). *A Theory of Justice*. Cambridge, MA: Harvard University Press.

Reisch, G. (1991). 'Did Kuhn Kill Logical Positivism?' *Philosophy of Science* 58: 264–77.

Ricoeur, P. (1970). *Freud and Philosophy*. (Orig. 1965). New Haven, CT: Yale University Press.

Rorty, R. (1979). *Philosophy and the Mirror of Nature*. Princeton: Princeton University Press.

Ross, A., ed. (1996). *Science Wars*. Durham, NC: Duke University Press.

Ross, D. (1991). *The Origins of American Social Science*. Cambridge, UK: Cambridge University Press.

Rothschild, E. (2002). *Economic Sentiments*. Cambridge, MA: Harvard University Press.

Rushkoff, D. (2010). *Program or Be Programmed*. New York: O/R Books.

Schiappa, E. (1990a). 'Did Plato Coin *Rhetorike?' American Journal of Philology* 111: 457–70.

———. (1990b). 'Neo-Sophistic Rhetorical Criticism or the Historical Reconstruction of Sophistic Doctrines?' *Philosophy and Rhetoric* 23: 192–217.

Schiff, T. (2006). 'Know It All: Can Wikipedia Conquer Expertise?' *New Yorker*, 31 July.

Schomberg, R. v. (2006). 'The Precautionary Principle and Its Normative Challenges'. In *Implementing the Precautionary Principle: Perspectives and Prospects*, edited by E. Fisher, J. Jones and R. von Schomberg, 19–42. Cheltenham: Edward Elgar.

———. (2013). 'A Vision of Responsible Research and Innovation'. In *Responsible Innovation*, edited by R. Owen, M. Heintz and J. Bessant. 51–74. London: John Wiley.

Schumpeter, J. (1942). *Capitalism, Socialism and Democracy*. New York: Harper & Row.

Schutz, A. (1946). 'The Well-Informed Citizen: An Essay on the Social Distribution of Knowledge'. *Social Research* 13: 463–78.

Sellars, W. (1963). *Science, Perception and Reality*. London: Routledge and Kegan Paul.

Sennett, R. (1977). *The Fall of Public Man*. New York: Alfred Knopf.

Serres, M. and Latour, B. (1995). *Conversations on Science, Culture, and Time*. Ann Arbor: University of Michigan Press.

Shapin, S. and Schaffer, S. (1985). *Leviathan and the Air-Pump*. Princeton: Princeton University Press.

Shaplin, A. (2009). *The Tragedy of Thomas Hobbes*. London: Oberon.

Simon, H. (1947). *Administrative Behaviour*. New York: Macmillan.

———. (1981). *The Sciences of the Artificial*. 2nd ed. (Orig. 1969) Cambridge, MA: MIT Press.

Sismondo, S. (2017). 'Post-Truth?' *Social Studies of Science* 47(1): 3–6.

Smil, V. (2001). *Enriching the Earth*. Cambridge, MA: MIT Press.

Smolin, L. (2006). *The Trouble with Physics*. New York: Houghton Mifflin.

Stinchcombe, A. (1990). *Information and Organizations*. Berkeley: University of California Press.

Stokes, D. (1997). *Pasteur's Quadrant: Basic Science and Technological Innovation*. Washington, DC: Brookings Institution Press.

Strauss, L. (2000). *On Tyranny*. (Orig. 1948) Chicago: University of Chicago Press.

Sunstein, C. (2001). *Republic.com*. Princeton: Princeton University Press.

Swanson, D. (1986). 'Undiscovered Public Knowledge'. *Library Quarterly* 56(2): 103–18.

Taleb, N. N. (2012). *Antifragile*. London: Allen Lane.

Tetlock, P. (2003). 'Thinking the Unthinkable: Sacred Values and Taboo Cognitions.' *Trends in Cognitive Science* 7(7): 320–24.

———. (2005). *Expert Political Judgement: How Good Is It? How Can We Know?* Princeton: Princeton University Press.

Tetlock, P. and Gardner, D. (2015). *Superforecasting: The Art and Science of Prediction*. New York: Crown Publishers.

Thompson, M., Ellis, R. and Wildavsky, A. (1990). *Cultural Theory*. Boulder, CO: Westview Press.

Tkacz, N. (2015). *Wikipedia and the Politics of Openness*. Chicago: University of Chicago Press.

Toffler, A. (1970). *Future Shock*. New York: Random House.

Tollison, R., ed. (1985). *Smoking and Society: Toward a More Balanced Assessment*. Lanham, MD: Lexington Books.

Turner, S. (2003). *Liberal Democracy 3.0*. London: Sage.

————. (2010). *Explaining the Normative*. Cambridge, UK: Polity Press.

Turner, S. and Factor, R. (1994). *Max Weber: The Lawyer as Social Thinker*. London: Routledge.

Tuvel, R. (2017). 'In Defence of Transracialism'. *Hypatia* 32: 263–78.

Vaihinger, H. (1924). *The Philosophy of 'As If'*. (Orig. 1911). London: Routledge and Kegan Paul.

Wark, M. (2004). *A Hacker Manifesto*. Cambridge, MA: Harvard University Press.

Wendt, A. (2015). *Quantum Mind and Social Science*. Cambridge, UK: Cambridge University Press.

White, M. (1949). *Social Thought in America: The Revolt against Formalism*. New York: Viking Press.

Woolgar, S. (1991). 'Configuring the User: The Case of Usability Trials.' In *A Sociology of Monsters: Essays on Power, Technology and Domination*, edited by J. Law, 58–97. London: Routledge.

Wootton, D. (2006). *Bad Medicine: Doctors Doing Harm since Hippocrates*. Oxford: Oxford University Press.

Wuthnow, R. (1989). *Communities of Discourse: Ideology and Social Structure in the Reformation, the Enlightenment and European Socialism*. Cambridge, MA: Harvard University Press.

INDEX

Milton Keynes UK
Ingram Content Group UK Ltd.
UKHW040631210924
1771UKWH00029B/116